Scanner Radio Guide

Scanner Radio Guide

by Larry M. Barker

A DX/SWL Press Book

HighText
publications inc.

Solana Beach, CA

Printed in the United States of America.

Cover design and illustrations: Brian McMurdo, Ventana Studio, Valley Center, CA
Developmental editing: Maria Luchadora–Fuerte, Enid, OK
Production services: Greg Calvert, Artifax , San Diego, CA

ISBN: 1–878707–10–8
Library of Congress catalog number: 93–077213

HighText is a registered trademark of HighText Publications, Inc.

Contents

Introduction

FORGET ALL THE CORNY police dramas you may have seen on television or in the movies—nothing can equal the excitement of listening in as real cops engage in a "hot pursuit" of armed suspects down a busy highway. And no fictional melodrama about doctors is half as gripping as the conversations between paramedics and hospital staff as an accident victim is rushed to a hospital by ambulance.

About twenty years ago, a new type of radio receiver became available to the general public. This radio tuned the very high frequency (VHF) and ultra high frequency (UHF) channels used by police and fire departments, the military, commercial businesses, etc. What was exciting about this type of radio was how it tuned the VHF and UHF bands. Instead of receiving just a single frequency, these radios would rapidly tune through, or "scan," a set of frequencies and stop on any frequency where there was a signal present. These radios became known as "scanners," and kicked off a rapidly growing hobby of listening to two-way radio communications on the VHF and UHF radio bands. Today millions of people follow the latest happenings in their communities by using scanners.

There's much more to hear on a scanner than just your local police, fire, and ambulance services. You can hear businesses using two-way radio systems to pick up and deliver people and packages all over town. You can listen in as reporters and television news crews cover major events. Your local airport will be a beehive of activity, with airplanes arriving, departing, and being directed on runways by the control tower. The military is a big user of frequencies tuned by scanners, and some scanners will even let you listen to communications directly from the Space Shuttle. Ham radio operators use the VHF and UHF bands to talk to their friends around town and even to link their personal computers together by radio. A portable scanner will let you listen to the behind-the-scenes activity at concerts, sporting events, rallies, and other major public happenings. And, if the atmospheric conditions are right, you can sometimes hear communications from hundreds or even thousands of miles away on a scanner.

Despite the popularity of scanners and scanning, there has been surprisingly little information about scanning available for the non-technical "scan fan." There are several excellent frequency guides for different areas of the country, but there has been very little telling the new scanning hobbyist what can be heard on the VHF and UHF bands. . . or explaining the technical terms used to describe scanners. . . or telling how to increase the range and performance of a scanning receiver. That's the purpose of this book. Even if you're an experienced scanner user, there's probably quite a bit you can hear on your scanner that you don't know exists. Hopefully, this book can expand your scanning horizons.

Scanning can give you a whole new perspective on the dangers and challenges faced each day by the men and women of your local police and fire departments. It can also give a feel for the daily pulse and routine of various local agencies—such as animal control, sanitation, and public works—that make life a lot better for all of us. And after listening to your local airport's air traffic control channels, you'll never again be blasé about a take-off or landing when you're flying.

Scanning isn't just a way to open your ears to your community; it's really a way of opening your eyes to people and events around you. And, best of all, it's fun!

Larry M. Barker

CHAPTER ONE

Some Fundamentals

S CANNING USES TERMS AND CONCEPTS that may not be familiar to you, especially if you have no previous experience with or knowledge of two-way radio systems. In this chapter, we'll look at some of the basic information you need to know to get the most out of any scanner.

Frequencies and Scanner Bands

The frequencies of radio signals are measured in units known as *hertz*, in honor of the radio pioneer Heinrich Hertz. A hertz represents a single complete cycle of a radio wave. The hertz (abbreviated Hz) is too small a unit to be of practical use when indicating the frequency of a station, so the units kilohertz (abbreviated kHz) and megahertz (MHz) are used instead. One kHz is equal to 1000 Hz, and a MHz is equal to 1,000,000 Hz or—to put it another way—1000 kHz. Suppose your scanner is tuned to 42.88 MHz. That frequency could be also be expressed as 42,880 kHz or even 42,880,000 Hz. But to keep things simple, all scanner frequencies are expressed in MHz.

The frequencies covered by your scanner are in the very high frequency (VHF) and ultra high frequency (UHF) ranges of the radio spectrum. Frequencies below 540 kHz are known as *longwave*, and those frequencies are primarily used by aeronautical and maritime navigation beacons. The standard AM broadcasting band lies from 540 to 1600 kHz (soon to be expanded to 1700 kHz in North America). Frequencies from 1600 kHz to 30,000 kHz (30 MHz) are known as *shortwave*.

Most scanners tune only the frequencies above 30 MHz (although some cover frequencies down to 25 MHz). The range from 30 to 300 MHz is considered VHF, while frequencies from 300 to 3000 MHz are considered UHF. (Frequencies above 3000 MHz are called the super-high frequencies, or SHF.)

Most scanners do not tune all frequencies above 30 MHz. Instead, they tune segments known as *bands*. This is because a good chunk of the VHF and UHF frequencies are used for purposes other than two-way radio communications. Television broadcasting uses a lot of VHF and UHF frequency space, and the FM radio broadcasting band is from 88 to 108 MHz. Almost all scanners will cover at least three bands used for two-way communications—VHF Low (30 to 50 MHz), VHF High (150 to 175 MHz), and UHF (450 to 470 MHz, sometimes referred to as UHF Low). Some scanners will cover some additional frequencies above and/or below these ranges (such as 25 to 50 MHz). Many scanners also cover the UHF-T band (470 to 512 MHz). This band is shared with television channels 14 to 20, and may not be used in your area if you are within range of a television station on those channels. There are many other bands where two-way communications are conducted, and we'll take a look at them later in this book. However, the "Big Three"—VHF Low, VHF High, and UHF—are where most of the action is.

The frequencies used for two-way radio communications are organized into regularly spaced *channels*. These channels are spaced apart from each other by a fixed amount of frequency space to avoid interference between stations. On the VHF Low band, channels are typically spaced 20 kHz apart from each other (30.20, 30.40, 30.60, etc.). On the VHF High band, the usual spacing is 30 kHz for business users (such as taxicabs and delivery services) and 15 kHz apart for other users like police and fire departments. On the UHF band, channels are usually spaced 25 kHz apart. In some areas, a slightly different spacing arrangement might be used. If you already have a program-mable scanner, you may have noticed that it tunes different bands in different increments depending on the channel spacing normally used on that band.

There are several other popular bands covered by many scanners, such as the 108 to 136 MHz aeronautical band used by airports and airplanes and the 225 to 400 MHz band used by many United States and foreign military services for aviation. We'll discuss other bands and their channel spacing later in this book. For now, the important thing to remember is that stations aren't scattered at random across the bands tuned by a scanner, but instead follow a definite pattern. And if you're in the market for your first scanner, I strongly suggest that you get one that covers at least the VHF Low, VHF High, and UHF bands. It will let you hear many exciting and interesting communications, and will make an ideal back-up or second scanner if you later decide to get a more advanced model.

What Can You Hear on a Scanner?

There's much more than just your local police and fire departments to be heard on your scanner. Here are some of the types of stations you can listen to:

- *Police and fire departments*. Most people probably buy scanners so they can monitor the activities of their local law enforcement and fire departments. In addition to your local police, however, you can also monitor your state police and highway patrol as well as law enforcement agencies at the national level, such as the Federal Bureau of Investigation, Customs, and the Royal Canadian Mounted Police.

- *Special emergency*. These stations include ambulances, paramedics, and rescue services.

- *Motor carriers and transport.* Taxicabs, trucks, railroads, buses, and delivery services can be heard on your scanner.

- *Aeronautical.* Airplanes, airports, and air navigation beacons (that is, stations that just repeat their location continuously) can be heard in the 108 to 136 MHz band. 121.5 MHz is a channel set aside for emergency communications involving any private or commercial aircraft.

- *Amateur or "ham" radio.* Hams have the use of frequencies at 28 to 29.7 MHz, 50 to 54 MHz, 144 to 148 MHz, and 420 to 450 MHz that can be tuned by most scanners. There are other ham bands above 30 MHz that some scanners tune.

- *Marine.* Marine stations include stations on boats and shore stations (such as marinas and the U.S. and Canadian Coast Guards) that contact boats.

- *Military.* As mentioned earlier, the 225 to 400 MHz band is heavily used by U.S. and foreign military services for aviation (air forces, naval aviation, army airborne services, etc.). However, there are also several channels in VHF Low, VHF High, and UHF bands used by the various U.S. military services. In fact, the U.S. military might pop up almost anywhere above 30 MHz. You don't have to be near a major base to hear activity on those channels; local National Guard units can be heard there as well.

- *Federal government.* Civilian agencies of the federal governments of the United States and Canada are heavy users of the VHF and UHF bands. In the United States, you can hear the communications of virtually all cabinet-level departments (Department of State, Department of Commerce, etc.) in addition to specialized agencies such as the Bureau of Land Management, Environmental Protection Agency, and Internal Revenue Service.

- *State and local governments*. In addition to your city's police and fire departments, you can also hear other city government functions, such as animal control, maintenance, schools, and public works, on your scanner. The same goes for various state agencies.

- *Private business*. Businesses are heavy users of the VHF and UHF bands. In the United States, the Federal Communications Commission (FCC) has established special radio services for such industries as motion pictures, petroleum, manufacturing, and forest products and logging. Other businesses running the gamut from construction work crews to amusement parks can be heard on your scanner.

- *Press*. Members of the print and broadcast media use the VHF and UHF bands to keep in touch with their offices, coordinate their activities when covering a major event, and even to transmit reports back to their offices.

- *General Mobile Radio Service*. This is a radio service in the United States open to any citizen or organization needing two-way radio communications. This service is much like a highly improved version of the citizens band (CB), and in fact this service was once known as the "Class A" CB band.

- *Paging*. Those "beepers" so many people carry with them receive signals transmitted on the VHF and UHF bands.

- *Mobile telephone*. Scanning has attracted some unfavorable—and largely erroneous—publicity in recent years because some scanners cover frequencies used by cellular telephones. (An ordinary television set can also let you hear cellular telephone calls, but more about that later.) However, scanners have for years covered the old mobile telephone channels in the VHF Low, VHF High, and UHF bands. Most scanners can also tune the cordless telephone frequencies around 46 and 49 MHz.

- *Walkie-talkies.* Those small unlicensed walkie-talkies so popular with kids operate around 49 MHz.

- *Weather bulletins.* The National Weather Service has a national network of stations operating in the VHF High band on a continuous basis with the latest weather information for your area. In the event of severe weather such as violent thunderstorms or tornadoes, these broadcasts are interrupted by special emergency announcements.

- *Miscellaneous.* There are numerous other users of the VHF and UHF bands besides the major ones we've just noted.

These include such activities as wireless microphones, eavesdropping and surveillance devices, remote control systems, and even roadside emergency callboxes.

Figure 1-1: The Bearcat 210XLT scanner by Uniden is an ideal "first scanner" for exploring the world above 30 MHz.

How Two-Way Radio Systems Work

The main station in a radio system is known as the *base* station. Examples of base stations are police headquarters, fire stations, or the headquarters of a business. Base stations are known as *fixed* stations, since they're located at permanent locations and don't move about. All stations that do move about— whether in cars, boats, airplanes, or as "walkie-talkies" carried by individuals—are called *mobile* stations. "Walkies-talkies" and other hand-held units are sometimes called *portable* stations.

The simplest two-way radio system consists of a base station and at least one mobile unit that communicate with each other over a single radio frequency (or channel). The base station and the mobile units take turns transmitting and receiving on the same frequency, and this is called *simplex* operation. The base station has the best antenna system and uses the highest transmitter power, sometimes over 100 watts. By contrast, mobile units use less effective antennas and lower transmitter powers. For example, a typical mobile unit in a car may have a transmitter power of only 10 watts while hand-held units carried by individuals may have powers as low as one or two watts.

While a simplex arrangement is uncomplicated and easy to set up, it does have some disadvantages. Perhaps the most serious are the lack of range and areas—known as *dead spots*—where communications are not possible. The normal communications range at VHF and UHF is what is called "line of sight." This means that the base station can communicate with any station out to the optical horizon, as seen from the base station antenna, plus an additional 10% to 20% beyond the horizon depending on local terrain. The higher the base station antenna, the greater the operating area of the system; a base station on a mountain top will cover a much greater range than one in a valley. However, most simplex systems will have a limited range of only a few miles unless the base station is at a high elevation. Hills and metal structures (like tall buildings) can absorb radio signals, and if a mobile station is in a valley or canyon it can drop from the "line of sight" of the base station. Mobile units can communicate with each other only over a short distance, perhaps only a mile or two in some situations. The limited range for simplex systems is why they are mainly used to cover relatively small areas such as shopping malls, public parks, businesses, amusement parks, etc. Simplex systems are seldom used by public safety agencies such as the police and fire department unless you live in a small town.

To extend the coverage of a two-way radio system, a *repeater* station may be used. A repeater is a relay station in a favorable location for extended VHF/UHF coverage, such as on top of a mountain, tall building, or tall tower. Unlike simplex systems, repeater systems make use of two separate frequencies known as the *input* and *output* frequencies. All mobile units in a repeater system will transmit on the input frequency of the repeater and receive on the output frequency of the repeater. The input and output frequencies are spaced sufficiently apart in frequency so there is no interference between the two. For example, the input frequency of a repeater may be 172.175 MHz and the output may be 172.975 MHz. The mobile stations in the system will transmit on 172.175, and the repeater station will listen on that frequency. When the repeater hears a signal on 172.175, it will immediately relay it on 172.975, which is the frequency that all units in the system will be receiving on. Thus, mobile stations can communicate with each other "through the repeater" over distances that would be impossible over a simplex system. The base station in a repeater system may be linked to the repeater by wire (usually telephone lines) or radio. Much of the activity you will hear on the VHF and UHF bands will involve some form of repeater system.

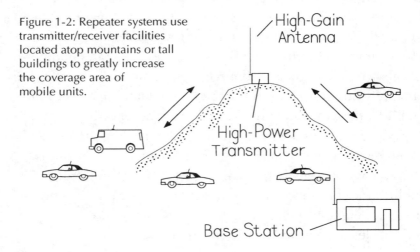

Figure 1-2: Repeater systems use transmitter/receiver facilities located atop mountains or tall buildings to greatly increase the coverage area of mobile units.

High-Gain Antenna

High-Power Transmitter

Base Station

Repeaters are popular with all users of the VHF and UHF bands. Most medium and large cities use repeaters for their police, fire, ambulance, and other public safety communications. Private businesses are also big users of repeaters, mainly through shared systems as explained below. Ham radio operators have set up their own repeater networks open to anyone with a ham radio license. And the military is a major user of repeaters, particularly at larger bases and facilities.

The input and output frequencies of a repeater are also known as a *channel*. The 172.175/172.975 MHz pair from our example is a single channel, and the channel is usually referenced by the output frequency, or 172.975 in our example. This method of transmitting and receiving on two different frequencies is called *half-duplex* operation.

Several different users can share the same repeater system by using some form of *coded access*. In a coded access repeater, each user of the system is assigned a special identifying code that is transmitted each time one of its units transmits. This code can be a set of tones (usually ones outside the range of normal hearing) or a special binary code like that used to transmit computer data. The repeater station will recognize only those codes assigned to authorized users of the repeater. If the repeater detects a signal on its input frequency with a correct access code, the repeater will relay the signal on its output frequency. Units listening on the output frequency will have their receivers activated when the access code matching theirs is relayed by the repeater. Several different users can share the repeater in relative peace under this system, since they will not receive transmissions not intended for them. Most large cities have several repeater systems which offer access to business users for a monthly fee, and each user is assigned an access code by the repeater owner.

A variation of the repeater system is *trunking*. While tone access helps cut down conflicts between users of a repeater, only one user can access the repeater at once. If another user needs to use the repeater, they must wait until the party currently using the repeater is finished. This can be a big problem if the users are public safety agencies such as the police or fire departments and delayed communications can literally be a matter of life or death. Trunking involves the use of more than one repeater and the automatic switching of communications to an unused repeater. The principles involved are similar to those used in your scanner, except instead of looking for busy channels a trunked repeater system looks for channels that aren't busy. When the user of a unit in a trunked repeater system first presses the microphone button, a burst of information is sent to the trunked repeater identifying the unit in use. The trunked repeater will find an unoccupied repeater frequency, and send back to the unit a burst of information telling the internal circuitry to switch to the designated repeater channel. The search for an unused repeater and the switching of units over to the repeater is handled electronically, and the entire process of locating an open channel and switching the unit to it usually takes only a fraction of a second. Trunking systems are becoming very popular for public safety communications in large urban areas, and most are found on frequencies above 800 MHz.

Both simplex and half-duplex communications allow only one side of a communication at a time; if the base transmits, the mobile units must listen and vice-versa. For a two-sided communication like a telephone call, *full duplex* operation must be used. Full duplex operation involves the use of different frequencies by the base station and mobile units. The base station receives on the mobile frequency and the mobile units receive on the base station frequency. Both the base station and mobile unit transmit continuously, allowing a normal back-and-forth conversation to take place. Full duplex operation makes mobile and cordless telephone service possible. If you have a cordless telephone, the base unit transmits on a frequency near 46 MHz while the cordless (mobile) unit transmits near 49 MHz.

Mobile telephone service today is provided by a form of trunking known as *cellular* radio. The mobile telephone service area is divided into several small cells. Each cell is a full duplex repeater covering a very limited area and connected to the conventional wired telephone system. At the heart of each cell is a computer controlling not only the cell itself but, in effect, the transmitting frequency of every mobile unit within the cell. When one of the mobile units within the cell transmits, the cell sends back a signal that "assigns" a channel (that is, a pair of frequencies) for the mobile unit to use. The various cells are linked together by a mobile telephone switching office (MTSO). If the mobile telephone moves from one cell area to another, the MTSO will switch the call from one cell to another and also change channels as required. If the signal from a mobile telephone is received by more than one cell, the MTSO will decide which cell will handle it. Cellular systems are found above 800 MHz.

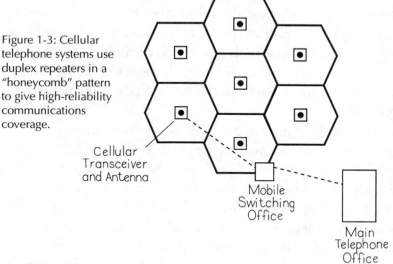

Figure 1-3: Cellular telephone systems use duplex repeaters in a "honeycomb" pattern to give high-reliability communications coverage.

Cellular Transceiver and Antenna

Mobile Switching Office

Main Telephone Office

There are other types of two-way radio systems in use above 30 MHz, but these are the major types. If you read that a certain network is a "trunked" system or "shares a repeater," you'll know what it means.

Is It Legal To Listen? Is It Fair To Listen?

A lot of controversy has arisen over scanners. Simply put, do users of two-way radio systems on the VHF and UHF bands have some expectation of communications privacy? This has become especially acute with the explosion of cellular telephone technology and many stories in the national news media about how cellular calls ". . . can be received by anyone with a scanner." As often happens where technical matters are concerned, the mainstream press has many of the facts incorrect.

The issue of communications privacy is not new. The Communications Act of 1934 established the Federal Communications Commission, and a provision of that act concerned privacy. Section 705 of that act prohibits using or divulging the contents of most non-broadcast radio transmissions, with certain exceptions such as ham radio communications or distress signals. This means that you cannot disclose the content or meaning of most signals you hear on your scanner—except mainly for ham radio signals and weather broadcasts—to anyone else. Moreover, under Section 705 you are prohibited from even telling a third party that any such communications took place. However, you are still free under Section 705 to listen to any communications as long as you do not tell anyone else what you hear.

As a practical matter, Section 705 has been largely and flagrantly violated for years. Communications on the VHF and UHF bands are routinely "used" and "divulged" by everyone from wrecker and towing services (that's how they appear so quickly at the scene of traffic accidents) to the press (that's how reporters appear so quickly at fires and crime scenes). A working scanner is part of the standard office equipment at many businesses and newspapers. However, there have been some isolated cases of prosecutions for violations of Section 705, usually in conjunction with divulging a communication to further a criminal activity (such as tipping off a drug dealer that the police have him under surveillance). Unless you're planning to do something equally stupid, you should have no problems with Section 705.

When cellular telephones became popular in the early 1980s, some users became alarmed to learn their conversations could be received by anyone with a receiver capable of tuning the 825 to 845 MHz range (used by the cellular telephones) and the 870 to 890 MHz range (used by the cells). To "protect" cellular users, cellular telephone interests used their well-compensated army of lobbyists and lawyers to ram the Electronic Communications Privacy Act (ECPA) through Congress in 1986. For the first time, this made it illegal to listen to certain types of radio signals in the United States. All mobile telephones and paging devices (but not cordless telephones) are off-limits, as are all signals that are encrypted or scrambled for privacy. The ECPA provides stiff fines and prison terms for violators. The ECPA does specifically exempt communications "readily accessible" to the public. Such communications are defined by the ECPA as including any branch of government (from local police to U.S. military), businesses, cordless telephones, ham radio operators, and citizens band radio. The provisions of Section 705 still apply under ECPA, by the way.

The ECPA was a hasty attempt to remedy a serious shortcoming in cellular technology, namely the lack of standard encrypting or scrambling. However, ECPA offers cellular users no meaningful privacy protection whatsoever. We already have plenty of existing laws concerning guns and drugs; the regularity with which they are violated shows how naive it is to expect a law alone to curb certain activities. Moreover, anyone deliberately violating the ECPA would almost certainly do so in private. How could any violation of the ECPA be detected unless the violator confessed?

Belatedly recognizing the basic futility of the ECPA, Congress in late 1992 passed a law outlawing the manufacture or sale of scanners capable of covering the cellular telephone bands. This particular act again demonstrates the dangerous level of technical illiteracy that permeates Congress, since virtually every home in the United States already has the capability to receive cellular telephone calls. The cellular telephone bands were carved out of television channels 70 through 84, and an ordinary television set can receive cellular telephone calls merely by connecting a UHF antenna. If you have a UHF antenna connected to your television (instead of cable), you can probably receive cellular telephone calls right now, especially if you live in an urban area. Try it—tune slowly upward from channel 70. You'll be able to hear calls on the audio of your television set, and the "snow" you see on the screen will form interesting patterns as the caller's voice changes. Like the ECPA, the 1992 "cellular scanner" ban gives cellular users the impression that something is being done to protect their privacy, but the true protection afforded will be negligible at best. The real solution—namely, to require all cellular telephones to include standard scrambling or encrypting circuitry—has been resisted by cellular telephone manufacturers eager to avoid any increase in their manufacturing costs. However, the switch of cellular telephones to digital and spread spectrum technologies will mean that within a few years it will be impossible to understand anything heard on the cellular frequencies on a scanner anyway. Hopefully, the arrival of that day will put an end to "scanner paranoia."

(An aside: the introduction of digital technology to cellular telephone service will make obsolete almost every cellular telephone now in use, requiring users to buy a new cellular phone. This is another interesting fact that cellular telephone companies are not telling their customers.)

The whole notion as to what degree of privacy a user of the radio spectrum is entitled to is one that has not been explored in the media horror stories about scanners and cellular telephones. Cellular telephone users—or all other users of the radio spectrum—do not in any sense "own" the frequencies they communicate over. Those frequencies, along with the rest of the radio spectrum in the United States, are legally the "property" of the people of the United States. The radio spectrum is, in effect, an electronic national park. Having a radio license or a cellular telephone is like having a camping permit for a national park; the license or permit allows you to use some of that space under specified conditions but does not convey permanent ownership. If you want privacy in a national park, put up a tent. If you want privacy on the airwaves, scramble or encode your signals. Otherwise, don't be too surprised if people see what you're doing or hear what you say—and don't rely on some largely unenforceable law to protect you.

There have been some scattered attempts by various state legislatures (such as California) and federal courts to extend privacy protection to cordless telephones. These have the same enforcement deficiencies as the ECPA.

An attempt was made in 1991 to basically outlaw most scanners through HR 1674, a bill passed by the House of Representatives. This bill would have outlawed the manufacturing or importing of any radio capable of ". . . receiving transmissions in the frequencies allocated to the domestic radio telecommunications service" or being readily altered to do so. The bill was sent to the Senate, where a strong lobbying effort killed it. The precise intent of this bill was difficult to fathom; were all of the millions of scanners already in existence supposed to immediately evaporate upon passage of the bill?

More troublesome have been the various state and local laws attempting to restrict the use of portable scanners or scanners installed in vehicles. Attempts by various local governments (particularly Philadelphia) in the 1950s to ban ownership of receivers capable of tuning police and fire departments were uniformly struck down by the courts, but some courts did permit banning of such receivers in automobiles.

The stated intent behind such laws is a desire to stop "ambulance chasing" and interference with police or fire department personnel at an emergency scene. However, the words of some public safety officials in support of such restrictions indicate their real motive is to keep the number of listeners to their communications to a minimum. (One suspects they would try to ban home listening if they could.)

At the time this book was written in 1993, the states of Indiana and Kentucky flatly prohibit all portable or mobile use of a scanner, although home listening is permitted. Florida has a curious prohibition on scanners at a "business premise" such as a towing service or newspaper. Rhode Island and South Dakota prohibit convicted felons of owning a scanner of any type. Minnesota, New Jersey, and New York prohibit "improper use" of a scanner, which is defined as using a scanner to aid in committing a crime or interfering with police, fire, or other public safety activities as a result of information obtained by listening to a scanner. Such restrictions on mobile or portable scanner use do change, and the dealer who sells you a portable or mobile scanner can advise you on any local restrictions on their use.

Residents of Canada are lucky. There are no restrictions whatsoever in Canada on what you may or may not listen to on a scanner, and no restrictions on portable or mobile use of a scanner either.

The best advice is to use discretion, courtesy, and common sense when using any type of scanner. Don't go repeating what you heard your neighbors say last night on their cordless telephone. Don't try to intentionally eavesdrop on cellular telephone calls. Don't show up at an emergency scene waving a portable scanner in the faces of police, fire, or other emergency personnel. And don't believe everything—or perhaps anything— that you read in newspapers or magazines about the threats to privacy posed by scanner users.

CHAPTER TWO

Choosing A Scanner: Understanding Features and Specifications

I f you're in the market for a new scanner, you might feel overwhelmed. Scanners have features often described in esoteric terms. The specifications used to measure scanner performance can likewise be bewildering. In this chapter, we'll take a look at scanner feature and "specs." We'll also discuss how to choose a scanner that best meets your needs and get the most performance for your scanner dollar.

Scanner Tuning Methods

Most scanners today are the *programmable* type. These have a keypad like a calculator or telephone that lets you enter frequencies you want to hear and assign them to a channel number (also known as a *memory*). You could enter your local police department frequency on channel 1, the highway patrol on channel 2, your local airport control tower on channel 3, and so forth. You can change the frequencies entered in each channel as you want. The frequencies you enter will be stored in your scanner in a form of electronic memory similar to that used to store information in a computer, and need some source of electric power to keep the frequencies stored. This is done as long as your scanner is connected to an AC wall outlet or has a fresh (fully charged) battery in it (even if the scanner is not turned on). Some scanners also have a "memory backup" feature that can retain the stored frequencies for a few hours or even days if it is no longer connected to an AC wall outlet or the battery is dead. However, once the stored frequencies are "lost" they will need to be entered again.

Some less expensive programmable scanners can only accept and tune frequencies in specific increments that vary from band to band. For example, a scanner might tune only in 5 kHz increments. If you try to enter a frequency such as 151.473 MHz, the scanner will instead round the frequency up to 151.475 kHz and accept that. In recent years, the Federal Communications Commission has been authorizing the use of "splinter" frequencies that end in a "half kilohertz," such as 462.5625 MHz (the last 5 represents 500 hertz). (Most users of splinter frequencies are low power hand-held portable units.) Scanners that tune in even one kilohertz increments can receive such splinter channels by tuning to the nearest whole kHz, but the frequency displayed by the scanner will be wrong. If the manual or specifications for a scanner doesn't indicate it can tune splinter channels, take a look at the frequency display of the scanner. If it can display four digits to the right of the decimal point, it can usually tune splinter channels. If it only displays three digits right of the decimal, it tunes in even kilohertz segments. Check the owner's manual for such scanners to determine if they can tune to each kilohertz or whether it tunes in increments such as 5 kHz.

Early scanners weren't programmable. You had to purchase a separate *crystal* for each channel you wanted to monitor. To monitor a new frequency, you had to buy and install a new crystal for that frequency. Today, almost all scanners are programmable types and "crystal-controlled" scanners are rare. However, there are a few "bargain basement" scanners that use crystals and crystal-controlled scanners show up at yard sales and flea markets. These should be avoided unless you're looking for a second scanner for a very specific purpose like listening only to your local police and fire channels. While crystal-controlled scanners are less expensive than programmable models (especially if they are bought used at a flea market), any savings can quickly be lost if you have to buy new crystals. The performance of crystal-controlled scanners is usually less than that of more modern programmable scanners.

Some scanners might be termed hybrids. They are programmable, but come with several frequencies (such as state police and highway patrol channels) permanently programmed into the scanner.

A few deluxe scanners are continuously tunable as well as

Figure 2-1: The Bearcat BC350A by Uniden is a hybrid scanner. It comes with several channels pre-programmed and you can program in additional channels used in your area.

programmable. Such scanners are sometimes described as having a *variable frequency oscillator* (VFO). These scanners can be tuned by turning a knob, just like a conventional AM or FM broadcast radio. Continuously tunable scanners are also continuous coverage; for example, one might cover 25 to 1000 MHz with no gaps. Such a scanner can tune the FM broadcast band and television channels as well as two-way radio frequencies.

What Types of Signals Can a Scanner Tune?

Not all radio signals found on the VHF and UHF bands are alike. Some are *frequency modulation* (FM) signals while others are some type of *amplitude modulation* (AM). Few scanners can tune every type of radio signal found above 30 MHz.

The most common signal type (known as a *mode*) is *narrowband FM*. Narrowband FM is similar to the FM used on the FM radio broadcasting band (88 to 108 MHz). However, narrowband FM occupies much less frequency space—about 10 kHz or less—than the frequency space occupied by an FM broadcasting station. This means narrowband FM doesn't give good fidelity for music, but works fine for voice communications. Virtually everything found on the VHF Low, VHF High, and UHF scanner bands will be in narrowband FM. Narrowband FM is also used for ham radio communications above 30 MHz. All scanners available today can tune narrowband FM, and this is the "default" receiving mode for scanners. If a scanner doesn't indicate which mode(s) it can tune, it's a safe bet that it receives only narrowband FM.

An FM signal that occupies more than about 10 kHz of frequency space is called a *wideband* FM (WBFM) signal. FM broadcast signals are all WBFM, but outside of that there aren't that many to be found. Perhaps the most notable are those used for voice communications by the Russian (formerly Soviet) manned space program. Some military signals are sometimes found in WBFM.

AM signals are like those found on the AM broadcasting band (540 to 1600 kHz). The big use for AM above 30 MHz is the 108 to 138 MHz aeronautical band. All communications on that band—including aircraft, control towers, beacons, etc.—are all in AM. AM is also used by the military aeronautical band that runs from 225 to 400 MHz. For scanners that cover the aeronautical band, you don't have to worry about changing from FM to AM when monitoring the aeronautical band, as the scanner will switch to the correct mode for you. The only other regular users of AM above 30 MHz are a handful of ham operators and some illegal, bootleg operators.

A type of AM is called *single sideband* (SSB). An AM signal consists of two identical sidebands and a carrier wave. SSB disposes of the carrier wave and one of the sidebands, and puts all of the transmitter energy into just one sideband. The result is a more efficient use of the transmitter power and greater range, but special receiving circuitry is required to properly receive a SSB signal. (On an ordinary AM or FM receiver, SSB sounds like Donald Duck after inhaling helium!) SSB signals are also more difficult to tune in properly. For these reasons, SSB is seldom used above 30 MHz except by some ham operators.

Unless you're an advanced scanner listener, you won't need to be concerned about any receiving modes other than narrowband FM and AM. The great majority (a good guess would be 95% or more) of the scanner bands activity outside the aeronautical band is narrowband FM, and the activity in the aeronautical band is exclusively AM. If you get seriously hooked on scanning as a hobby and don't want to miss out on any signals you might run across, then a "multimode" scanner can be a good investment for you. But if you're just starting out, an ordinary narrow-band FM (and AM, if it covers the aeronautical band) scanner will do fine.

Figure 2-2: The IC-R100 by Icom is a deluxe scanner for the world above 25 MHz. In addition to FM, it tunes AM and SSB with ease and scans up to 100 channels.

Sensitivity and Selectivity

These are two important but often misunderstood specifications for scanners. *Sensitivity* refers to how well a scanner can receive weak signals. *Selectivity* is how well the scanner can reject interfering signals near the frequency you want to receive. The relative importance of these two specifications depends on where your listening site is located. In some cases, you'll want more selectivity; in others, more sensitivity.

Sensitivity is measured in microvolts, abbreviated μV. A microvolt is one-millionth of a volt. When a signal is intercepted by your scanner's antenna, it produces a weak voltage that is delivered to your scanner. If a scanner has a sensitivity specification of 1 μV, it means that a signal producing a voltage of 1 μV in the antenna will also produce readable—although weak—audio from the scanner. In effect, sensitivity is a measure of the faintest signal that a scanner can hear. The lower the signal level in microvolts, the more sensitive the scanner is. A scanner with a sensitivity specification of 0.5 μV is more sensitive than one with a 1.0 μV specification.

Selectivity is rated in kHz at a certain decibel (dB) rejection level. The decibel is a ratio between two signal levels, and for selectivity specifications indicates the ratio between the desired signal and an interfering one on another frequency. A specification of 30 kHz at 50 dB (sometimes expressed as "– 50 dB" or "50 dB down") means signals located 30 kHz away from the one you want to receive are reduced in strength by 50 dB, which is equivalent to the desired signal being 100,000 times stronger than the interfering signal. The smaller the kHz value for 50 dB rejection, the better the selectivity because the scanner would be able to reject strong interfering signals closer to the desired one. One thing to look out for is the somewhat tricky way some scanner manufacturers give selectivity specifications. For example, a 30 kHz selectivity could be expressed as +/- 15 kHz. This is technically accurate, but misleads some people who don't pay attention to the + or - signs. A more sneaky trick is to use some rejection level less than the standard 50 dB point. This can really be misleading because decibels are logarithmic; they increase or decrease more rapidly than you might suspect. You might think at first that a 50 dB rejection level is only slightly better than a 40 dB rejection level, but in reality the 50 dB rejection level means a ten times better interference rejection than the 40 dB level.

The number of kHz for 50 dB rejection of an interfering signal is known as the *bandwidth*. The amount of bandwidth required varies by mode. FM signals (both narrowband and wideband) occupy more frequency space than AM and SSB signals, and require wider selectivity bandwidths. Typical bandwidths for 50 dB rejection offering good selectivity in most situations are 40 kHz for narrowband FM, 150 kHz for wideband FM, 6 to 10 kHz for AM, and 2.4 to 4 kHz for SSB.

If you're in a rural area, sensitivity is of more importance than selectivity. There are fewer stations in rural areas, and they tend to be located far apart in frequency. A very sensitive scanner lets you hear distant stations out to the maximum normal range (the horizon plus 10% or so extra) found on the VHF and UHF bands. In an urban area, by contrast, many channels will have several stations active. Being able to receive signals from far away is often not that important since many channels will have nearby stations active on them. The challenge instead is to be able to clearly separate channels from each other and prevent strong stations on adjacent channels from interfering with stations you want to hear. A scanner with "tighter" selectivity is what you need in an urban area. If you're located in a highly populated area, such as the tri-state area surrounding New York City or the Los Angeles basin, then a scanner with better than average selectivity—such as 30 kHz or less for 50 dB rejection—may be a good investment for you.

In general, more expensive scanners have better sensitivity and selectivity specifications than cheaper scanners. A big factor in the effective sensitivity of a scanner is the antenna used. The more signal an antenna delivers to a scanner, the more benefit you will get out of your scanner's rated sensitivity. An outdoor antenna or a preamplifier device between your scanner and its antenna can make a big improvement in the number of faint signals you can hear. Unfortunately, there is no simple way to improve the selectivity of your scanner since some internal circuit modifications are inevitably required. If you are troubled by interference from nearby frequencies, a directional "beam" antenna (which we'll discuss in the next chapter) can sometimes help by reducing signal pick-up from the interfering station while maximizing reception from the desired station. Otherwise, there's not too much you can do about inadequate selectivity outside of buying a better scanner.

"Intermod," Images, and Birdies

"Intermod" is short for *intermodulation*, a form of interference that is produced within a scanner when two strong signals mix together in the scanner's circuitry. The result is that one signal is superimposed on another or you hear bits and pieces of two or more signals on a single channel.

Intermod can be a very frustrating condition, since it's produced by strong signals in your area. The same scanner that has severe intermod problems in the city might have no intermod problems in the country. The ability of a scanner to handle strong signals without producing intermod or other problems is called the *dynamic range* of a scanner. Dynamic range is measured in decibels (dB), and the higher the number of dB the better the dynamic range of the scanner. Unfortunately, this specification is almost never given in scanner advertisements or owner manuals; about the only place you ever see it is in technical reviews of scanners in radio and electronics magazines. The only good guideline to the dynamic range of a receiver is its price—the more expensive the scanner, the better its dynamic range tends to be. (However, a few expensive scanners have mediocre dynamic range.)

Images are phantom signals in a scanner that are identical to a signal received by a scanner, except that images are on different frequencies than the actual signal. Images can be a problem by fooling you into thinking the image is an actual frequency and by causing interference to signals legitimately on the same frequency as the image. Images are produced by the internal signal conversion circuits used by all scanners, and all scanners will have at least a few images. However, on better scanners these images will either be too weak to be noticed or will lie outside of the tuning range of the scanner. *Image rejection* (also known as *spurious rejection*) is the measure of how well a scanner minimizes problems from images. It is also measured in decibels, and a figure of 50 dB or greater is considered excellent. Unfortunately, many scanner manufacturers do not specify an image rejection specification for their units. In these cases, the same advice mentioned earlier still applies: the more expensive the scanner, the better the image rejection tends to be.

Birdies are also produced in the internal signal conversion circuits in a scanner, but can be present even if no signal is being received from the antenna. If you come across a birdie on a channel, you will hear what sounds like a signal with no audio (that is, it sounds like somebody holding the microphone button down but saying nothing) or just random noise. Depending upon the scanner design, you may be able to eliminate birdies by setting the scanner's squelch control ("squelch" will be defined later in this chapter) to a high level. To determine if a signal on a scanner is a birdie or not, disconnect the antenna from the scanner. If you can still hear the signal with no apparent loss in signal strength, the signal is a birdie.

Features and Controls

Scanners today are loaded with different features and controls to make them easier to use and your listening more enjoyable. Here are some of the more common ones and the benefits they offer:

- *Squelch*. This is one control your scanner is certain to have. If you scan past a channel with no signal on it, you'll hear random background noise. This would be really annoying for monitoring, so scanners include a squelch control to silence the scanner when no signal is being received. To use your scanner's squelch, adjust it to the point where the background noise disappears. This is called the squelch *threshold*. When a signal is received, it "breaks" the squelch and you'll hear the signal. When you increase the setting of the squelch beyond the threshold point, a stronger received signal will be necessary to break the squelch.

- *Delay*. There's often a pause or delay between transmissions on a channel. For example, there is usually an interval between a call by a base station and the reply from a mobile unit. If the interval is too great, the scanner will resume scanning before the reply is made. A delay feature (also known as scan delay) will cause the scanner to wait on a channel for a few seconds after the last transmission so you can hear any late replies. This feature can be switched on or off.

- *Banks*. Several scanners today can scan 100 channels or more. To help you better organize and control so many channels, these scanners usually let you divide coverage into different "banks" of channels. For example, a 100 channel scanner might let you group channels into five banks of 20 channels. You can place all police and fire channels into one bank, all aeronautical channels into another, all ham radio frequencies into a third bank, and so forth. You can select which banks you want to scan. This is handy during emergency situations; for example, you might only want to monitor police and fire channels. You could choose that bank and ignore the others, thus reducing the chance of missing a call as the scanner scans through the other banks. Some scanners allocate certain banks for scanning and others for use with the search function (see below).

- *Priority Channel.* There might be a certain channel where you never want to miss a call, such as your local police or ambulance frequency. This feature lets you designate such a channel. Your scanner will check the priority channel at intervals of a second or two. Whenever a signal appears on that channel, your scanner will immediately switch over to it regardless of which channel or bank it is currently scanning. Some scanners have two or more priority channels.

- *Lockout.* Some channels may be temporarily busy with activity that you do not want to monitor. This lets you "lock out" those channels from the scanning sequence without deleting them from your scanner's memory. You can just as easily "unlock" such channels when you want to monitor them again.

- *Search.* This is a really handy feature for finding new channels active in your area. You specify the upper and lower frequency limits of a search, and your scanner will tune through that range in specified frequency increments (like 5 kHz) and stop whenever it finds a frequency in use. The frequency will be displayed on your scanner's frequency readout, so you can note it and add it to your scanner if you wish. A lot of scanner fans love nothing better than finding a new frequency being used in their area!

Figure 2-3: Advanced portable scanners, like the PRO-43 from Radio Shack, give the features and performance of base scanners in a convenient hand-held size.

- *Autoload.* This feature is sometimes called "Monitor" or other term, and allows active frequencies discovered during a search to be automatically loaded into unprogrammed channels on the scanner by pressing a button.

- *Hold.* This feature lets you stop a scan or search at a desired frequency by pressing a button. You can then monitor the frequency continuously.

- *Attenuator.* This controls lets you make a moderate reduction in the sensitivity of your scanner. This can be handy if you're experiencing intermod or other problems caused by too many strong signals in your area.

- *Recording Output.* Many listeners like to record transmissions they intercept, and a recording output lets you do just that. A voice-activated tape recorder is especially useful with this feature, as the squelch will keep the scanner silent (and the recorder off) unless a signal is received.

- *Computer Interface.* A handful of deluxe scanners allow you to control their operation using a personal computer. The control software used with the computer lets you monitor certain channels at different times and to conduct complex search routines.

- *Weather Channels.* Several scanners allow you to immediately access the frequencies used by the National Weather Service for their continuous weather broadcasts on the VHF High band. The frequencies are pre-programmed into the scanner.

- *Signal Strength Meter.* This is another feature usually found only on deluxe scanners. This is a meter that gives a visual indication of the relative strength of different signals received.

- *Hyperscan or Turboscan.* This feature might go by other names on different scanners, but it is simply an optional high speed scanning rate (such as 50 channels per second on some scanners).

- *Sound or Audio Squelch.* Some signals found on the VHF and UHF bands may be continuous but have no audio. The most common are from room monitor units when no one is in the room. Sometimes police or fire mobile units get their microphone button stuck in the "on" position when no one is using the unit, and repeater stations can also get stuck in "on" when they're not relaying a mobile station. A signal will still be transmitted, but you won't hear any audio. The signal is still capable of stopping a scanner on its frequency, however. To prevent this, sound squelch will resume scanning if no audio is present on a received signal within a few seconds after stopping on a frequency. Sound squelch ignores "birdie" signals.

Selecting Your First Scanner

There's no right scanner for everybody. What might be a good choice for you might be a bad choice for someone else—the key is what your interests are and how your scanner might be used.

If you're looking for your first scanner, your choices will probably come down to either a hand-held portable unit or a base (powered from the AC wall socket) unit. Most hand-held units can be used at home by using an AC adapter to power the unit instead of batteries, and it's possible to connect base unit antennas to most hand-helds. As a general rule, hand-helds are slightly more expensive than base units with the same features, number of channels, and performance specifications, although the price differential is usually not significant enough to be a major factor in a purchase decision. If you do think you'd like to monitor the behind-the-scenes activity at sporting and other public events, then a hand-held scanner makes a great choice for a first scanner. (Earphones will let you listen privately in public.) If you just want to listen at home, then a base scanner will usually give you a little more bang for your buck.

Your first scanner should cover at least the VHF Low, VHF High, and UHF bands. These will have plenty of activity regardless of where you live and offer a good sample of what you can hear on most other bands. The big exception is that you will not hear any aeronautical activity. For that, you will need a scanner covering the civilian aeronautical band (108 to 136 MHz). Aeronautical listening isn't for everyone's taste, and some big fans of the "land mobile" channels couldn't care less about the "aeronautical mobile" bands. But there's no way to know if you'll like the aeronautical bands if you don't try them. Most scanners covering the civilian aeronautical band are about 10–20% more expensive than comparable scanners without such coverage. Unless price is a crucial factor in your purchase decision, I suggest that your first scanner should include aeronautical coverage.

The number of channels you will need will depend on where you live and what communications you are interested in hearing. In a rural area, 10 channels might be plenty, but in a major city 40 might not be enough. I would suggest that your first scanner have at least 20 channels if you live in an urban or suburban area. You might get by with 10 channels if you live in a rural area, or are only interested in exclusively monitoring certain communications such as your local police and fire departments. However, the first thing most scanner fans wish for is a scanner with more channels, so the more channels you can afford on your scanner the happier you will probably be in the long run.

One feature that you will probably appreciate a great deal is a search function. While there are several excellent frequency guides and directories available for the VHF and UHF bands, new stations are always coming on the air and existing ones change frequencies. Moreover, some of the most interesting listening involves so-called "itinerant" users of the bands. Itinerant users are sporting events, concerts, political campaigns, evangelical crusades, parades, and other short-term activities that move from city to city or happen rarely (like parades). Without some sort of search function to locate frequencies used by itinerant users, you will miss out on a lot of fascinating and often unusual communications. Scanners with a search function are more expensive than those without, but you will probably find the extra cost to be a wise investment.

Other features most scanner listeners appreciate are at least one priority channel and channel lockout. I have the priority channels in my scanner set to the international aviation emergency frequency (121.500 MHz) and the main channel used by my local police department. I have frequencies for some local ham radio repeaters entered into my scanner, and the lockout feature is handy when some of the ham users get long-winded and prevent my scanner from searching other frequencies. Whether you need the other features covered earlier in this chapter is up to you and your preferences.

For a first scanner, you will only need to receive narrow-band FM (and AM if you get a scanner with aeronautical band coverage). The other modes (wideband FM, AM outside the aeronautical bands, and SSB) make up a small percentage of the activity above 30 MHz, and you won't be missing much if your first scanner doesn't receive them.

How much should a first scanner with the features and capabilities we've just discussed cost? That depends on where you shop and the exact model you buy, but you should be able to find an adequate first scanner for about $200.00 (U.S.) or less.

Antennas

Most scanners come with some sort of antenna as standard equipment. Base units will have a telescoping "whip" antenna that connects directly to the back of the scanner, while hand-held portable units will have a short, flexible antenna known informally as a "rubber ducky." These simple antennas might be all you ever need and, depending on your circumstances, they might even be all the antenna you should ever use. But in most cases you can greatly increase the number of stations you can hear by using an outside antenna specially designed for scanner reception.

One thing you'll soon notice about all scanner antennas, whether the one supplied with your scanner or a deluxe out-door model, is that all are *vertical* antennas. This means all such antennas are installed and used while physically in a vertical position. If such antennas are placed in a horizontal position, the strength of the signals you receive will drop significantly. Like light, radio waves on the VHF and UHF bands are "polarized," and the position of the transmitting antenna determines the polarization of the radio wave. A horizontal antenna will be "blind" to a vertically polarized radio wave, and a vertical antenna will likewise be blind to a horizontally polarized radio wave. This effect gets more pronounced as the frequency increases; signals at 800 MHz are much more sensitive to polarization than those at 30 MHz. The reason why vertical polarization is used on the scanner bands is because it is easier to install a vertical antenna on a car or hand-held portable unit than a horizontal antenna. (Just imagine driving a car down the highway with a horizontal antenna sticking out from your car into the next lane!) For this reason, vertical polarization rules above 30 MHz and antennas must also be vertical.

The telescoping whip and "rubber ducky" antennas supplied with most scanners will generally do fine for local (within approximately 30 miles or so of your location) reception. If you're an urban listener surrounded by numerous stations, trying to use an outside antenna with a less expensive base or portable scanner can be counter-productive. An outside antenna will deliver more signal power to the scanner, but the scanner might be unable to handle the increased signal levels and intermod could result. Since many two-way radio systems in urban areas use repeaters located atop hills or tall buildings to improve coverage, an indoor antenna will generally do a fine job of pulling in most activity in your local area.

If you live in a rural or suburban area, or if you have a better quality scanner with a good dynamic range specification, then you should consider an outdoor antenna. Most outdoor scanner antennas are *omnidirectional*, meaning they receive signals equally well from all directions. Most are also "broadband," covering wide frequency ranges from 25 to 1300 MHz. Perhaps the most popular omnidirectional scanner antenna is a type known as the *discone*.

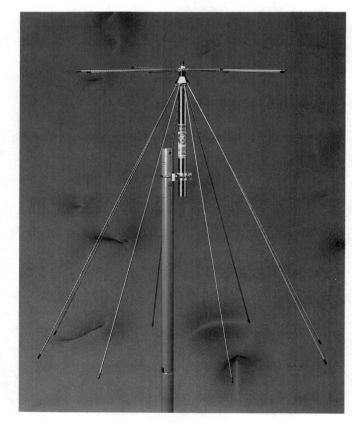

Figure 2-4: The discone is an excellent choice for an omnidirectional outdoor antenna.

It is also possible to use directional antennas, commonly known as *beam* antennas, with your scanner. A beam antenna will receive signals well in one direction but reject signals coming from other directions. Signals coming from the favored direction will also sound louder than if received using an omnidirectional antenna. This effect is known as *gain*. This combination of gain and rejection of signals from directions other than the desired one make beams a favorite choice of serious scanner hobbyists. They are especially useful in urban areas to cut down on intermod (due to rejection of unwanted signals) or to consistently receive signals from distant points (often up to 200 miles depending on terrain). There are a couple of beam antennas specifically designed for scanner reception available. To receive signals from different directions, you can use a television antenna rotor to turn the beam. Directional Yagi-type outdoor television antennas for channels 2 through 13 can also be used as beams on the VHF Low and VHF High bands, although their performance will not be as good as a beam designed especially for scanner reception.

Since normal reception range at VHF and UHF is slightly beyond the optical horizon, you can increase your reception area by getting the antenna higher in the air. This is true whether you're using an omnidirectional or beam antenna. Many serious scanner listeners have their outdoor antennas installed on a telescoping mast like those used for television antennas. Also, a difference of a few feet in where the antenna is located horizontally can make a big difference in reception of distant stations at frequencies above approximately 400 MHz. This is because distant obstructions, such as hills and tall buildings, can effectively block a signal at UHF. Moving the antenna slightly can mean a clearer path for the signal to reach your antenna. It's often worthwhile to temporarily place an outdoor antenna in different locations to determine which one gives you the best reception of the stations and channels you're interested in, and then permanently install the antenna in the location that gives the best results.

When installing any outdoor scanner antenna, follow the manufacturer's directions carefully. In particular, avoid placing the antenna where it could come into contact with power lines by any means whatsoever (such as falling across one). Another problem is lightning, and a device known as a lightning arrestor should be installed between your antenna and scanner to route lightning strokes to ground. The installation directions for your antenna will usually have some safety and lightning protection guidelines that you should follow precisely.

Scanners designed especially for use in a car or other vehicle do not generally come with an antenna and you'll have to buy one. Perhaps the most popular types are window glass-mounted types similar to those used for cellular telephones. In fact, these antennas are designed to look like cellular antennas so it's not obvious that you have a scanner in your car. Like cellular antennas, these install on the back window glass without the need for drilling any holes or other alterations to your car.

If you install an outdoor antenna, you will need *coaxial cable* ("coax") to connect the antenna to your scanner. Some antennas come supplied with a coaxial cable, and that should be used whenever possible. If you must supply your coax, don't use the same type of coax that you might use with a citizens band (CB) radio or a cable television system. Instead, you will need to use a special "low loss" cable designed for use at VHF and UHF frequencies. If you use the sort of coax that might be appropriate for a CB radio (like the RG-58 variety), a lot of the signal energy that your antenna intercepts will be dissipated in the coaxial cable before it reaches your scanner. To prevent this, use a premium quality cable such as RG8X, RG8U (also known as 9913), or RG188. The directions that come with your scanner antenna will often recommend a specific type of coax, and you should follow the manufacturer's advice. All coaxial cables have some signal loss at VHF and UHF, so keep the length of the cable between your antenna and scanner as short as possible.

The loss of any coaxial cable increases as the frequency of the signal it carries increases. Coax has a much higher loss at 800 MHz than it does at 40 MHz, for example. To compensate, a *preamplifier* can be installed between the antenna and scanner. The preamplifier will amplify the signals received by the antenna, and is usually installed at the antenna to amplify signals before they get sent down the coaxial cable. If you're located in a rural or other "weak signal" area, a preamplifier can be valuable. However, preamplifiers must be used with care. Preamplifiers amplify strong signals as well as weak ones, and can cause intermod in your scanner. The best advice is to try to get the most performance out of your antenna, or switch to a better model, before adding a preamplifier.

One possible solution to intermod and other interference problems is to add a *notch* or *bandstop* filter between your antenna and scanner. These are commonly known just as "scanner filters," and can be tuned to reject a specific frequency. If you are suffering from intermod or interference from an adjacent channel, you can tune the filter to reject the frequency of the interfering signal. Such filters come with coaxial cable connectors on either end, and it's simple to install them either at the antenna or your scanner. If intermod is a serious problem for you, a notch filter may be your best solution.

CHAPTER THREE

Propagation

PROPAGATION IS THE TERM used to describe how a signal travels from a transmitting station to a radio receiver. On the scanner bands, there are several different forms of propagation. As we've seen in earlier chapters, the typical range of a signal on the VHF and UHF bands is slightly greater than the horizon as seen from the transmitting antenna. But there are several times each year when special forms of propagation make it possible to receive scanner signals from up to about 1000 miles away. And, during years of high sunspot numbers, it's possible to listen to stations from the other side of the world on the 30 to 50 MHz band. Best of all, scanner reception over such distances requires no special equipment or antennas, only a knowledge of propagation.

Normal Propagation

As we discussed earlier, most propagation above 30 MHz is line-of-sight.; that is, from the transmitting (or receiving) antenna out to the horizon as seen from the antenna plus a slight additional amount beyond the horizon. This "bonus" coverage beyond the horizon is produced by the Earth's atmosphere, which "bends" the radio signals over the horizon for a short distance. Eventually, however, radio signals at VHF and UHF will continue traveling away from the Earth's surface beyond the horizon. The signals will move through the upper layers of the atmosphere and be lost into outer space. The normal coverage range of signals above 30 MHz will vary according to several factors, such as the terrain in your area and the height and location of the transmitting and/or receiving antennas. Another important factor is the frequency of the signal. Signals on the VHF Low band can normally be reliably heard about 20% to 35% further away than a signal of the same transmitter power and antenna height on UHF. However, the use of repeater stations and trunked repeaters can give UHF networks the same coverage as VHF systems.

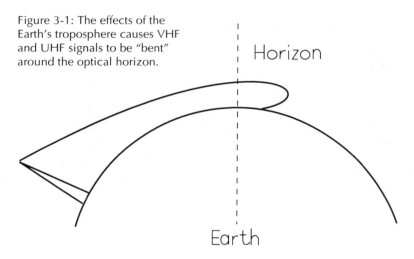

Figure 3-1: The effects of the Earth's troposphere causes VHF and UHF signals to be "bent" around the optical horizon.

Horizon

Earth

Such variables make it difficult to generalize about what signal coverage range you should expect. However, most scanner listeners can hear signals within a 20 to 30 mile radius if they are using an indoor antenna or a portable hand-held scanner, and within a 50 to 75 mile radius if they are using an outdoor antenna or are in a favorable location such as a hilltop. These generalized coverage areas are based on the VHF High band, and will therefore be greater on VHF Low but less on UHF.

One of the advantages of the VHF and UHF scanner bands is that their normal range is fairly constant regardless of time of day, weather, or season of the year. Most disturbances or propagation irregularities are the result of your local terrain. We've already mentioned one such disturbance, known as dead spots.

Dead spots are especially common in hilly or mountainous areas, and are all too familiar for scanner listeners in cities like Los Angeles and San Francisco. As mobile units travel, their signals may be blocked by hills or mountains. Tall buildings in metropolitan areas can produce the same effect. Another problem is called *picket fencing*. This is found on signals from mobile units in moving cars, and sounds like a rapid "flutter" of the signal. This is caused by the signal being temporarily blocked by buildings and other obstructions the car passes while in motion.

At frequencies above 400 MHz or so, an interesting new phenomenon comes into play. Signals at such frequencies can be absorbed somewhat by the foliage on trees, creating dead spots and picket fencing in heavily forested areas. If you live in an area where trees lose their leaves each fall and regain them in the spring, one of the sure signs of spring in your area could be the annual reappearance of some dead spots on UHF that had vanished the previous autumn!

Tropospheric Bending

The troposphere is the layer of the atmosphere that lies closest to the Earth, and is the layer where weather takes place. It is the troposphere that is responsible for bending signals slightly over the horizon in normal propagation. However, it often happens that weather conditions can "trap" VHF and UHF signals and keep bending them over the horizon for distances up to several hundreds of miles before they finally escape into space. This is known as *tropospheric bending*, often referred to as "tropo" by scanner fans.

Tropospheric bending is caused by temperature inversions. Normally, air temperature drops with elevation. However, under unusual weather conditions, the air temperature may actually start to increase at a certain elevation. The usual cause of this happens along the boundaries of occluded or stationary weather fronts, where a mass of cool, drier air meets a mass of warmer, moist area. The cool, drier air can sometimes force its way "under" the warm, moist air and push it upward. Clues that a temperature inversion is taking place include fog, high haze, and smog. When a VHF or UHF radio signal enters a temperature inversion, signals continue to be bent beyond the horizon close to the Earth for the length of the inversion. In continental North America, VHF and UHF signals have been received up to 1400 miles away by tropospheric bending. The record is over 2500 miles, which has been recorded several times between California and Hawaii. Most "tropo" propagation covers only a few hundreds of miles, however.

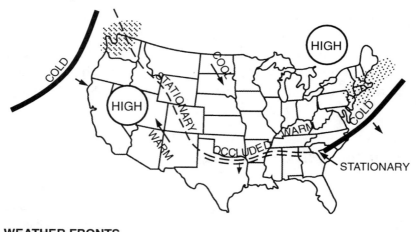

WEATHER FRONTS

WARM	════════	
COLD	▰▰▰▰▰▰	
OCCLUDED	═ ═ ═ ═	
STATIONARY	─ ─ ─ ─	

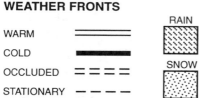

RAIN

SNOW

Figure 3-2: Weather maps such as this can be good indicators that enhanced reception due to tropospheric bending is possible. In this case, "tropo" may be possible along a path from the Texas panhandle to the Carolinas.

    Tropospheric bending is very common along the Atlantic, Pacific, and Gulf of Mexico coasts during mid to late summer. It happens at sunset, when cool breezes from the water blow inland and force the warm inland air upward. At such times, scanner listeners located near the Atlantic coast can hear VHF and UHF stations from southern Florida to the Canadian maritime provinces. Tropospheric bending is very common across the Gulf of Mexico in summer, with scanner listeners in Tampa, New Orleans, and Houston being regular listeners to each others' businesses and public service agencies. Late summer smog alerts in California are the result of temperature inversions, and experienced scanner listeners there know a smog alert also means stations from San Diego to San Francisco can be heard.

Further inland, tropospheric bending happens mainly in the fall when occluded and stationary weather fronts are common. Most "tropo" propagation happens east of the Rocky Mountains, as the Rockies disrupt temperature inversions. (The Appalachian Mountains are not high enough to cause such problems.) If you live in states like Utah, Idaho, or Montana, "tropo" will be rare. In fact, only the Pacific Coast experiences significant "tropo" west of the Continental Divide.

A short-lived form of tropospheric bending can happen around sunrise almost any time of the year during clear or foggy weather. At sunrise, the upper part of the troposphere often gets warmed before the lower part, producing a temperature inversion that might last for 20 to 45 minutes after sunrise. This form of "tropo" usually has a range of less than 300 miles.

The effects of tropospheric bending are more pronounced as the signal frequency increases; signals on the UHF and VHF High bands are more likely to be propagated by tropospheric bending than those on the VHF Low band. Signals propagated by tropospheric bending tend to be less strong than those from your local stations. Tropospheric bending can last for several consecutive days, and signals tend to gradually fade out over a period of several minutes when the temperature inversion finally breaks down.

No special scanner or antenna is necessary to receive signals via tropospheric bending, although an outdoor antenna will certainly help you hear more "tropo." A directional beam antenna is especially valuable, as it lets you get maximum signal reception from the direction of the temperature inversion.

Sporadic-E Propagation

The Earth's atmosphere has parts above the troposphere. One of them is the *ionosphere*, which lies about 35 to over 400 miles above the Earth's surface. The ionosphere is named for the numerous electrically charged particles, called *ions*, which are found there. Ions get their electric charge from the Sun's radiation. The ions in the ionosphere form a reflective "mirror" for radio signals below 30 MHz. Signals on such frequencies are bent back to Earth by the ionosphere instead of traveling out into space. The charged particles of the ionosphere make international communications possible on the shortwave radio bands, and are also the reason why stations from hundreds of miles away can be heard on the AM broadcast band (540 to 1600 kHz) at night. Normally, signals above 30 MHz pass through the ionosphere without being bent back to Earth. This is because the level of ionization in the ionosphere is not heavy enough to reflect VHF and UHF signals back to Earth. However, there are some ways the level of ionization can rise to a sufficiently high level to reflect VHF and UHF signals. These are collectively called "skip," and one of them is known as *sporadic-E*, sometimes called "E-skip."

Sporadic-E happens in the "E" layer of the ionosphere, which is found between 35 to 60 miles above the surface of the Earth. At certain times of the year, "patches" or "clouds" of very intense ionization form in the E layer, and these are indeed capable of reflecting signals above 30 MHz back to Earth so they can be received up to 1500 miles away. The formation, movement, and duration of these clouds are very erratic, which is why this phenomenon is called "sporadic." The exact cause of sporadic-E is not yet known.

Sporadic-E happens in two well-defined "seasons" in the northern hemisphere. The major one runs from late May to early August, with a peak in late June and early July. The secondary one runs from early December to early January, with most activity centered around the winter solstice. Most sporadic-E happens from approximately 10:00 a.m. to noon (local time at your listening location) and again from about 4:00 to 10:00 p.m. However, sporadic-E can happen at any time of day or any time of year.

The ionized sporadic-E clouds move in the ionosphere. This means that you might first notice stations from a few hundred miles to the northeast of you coming in on a channel. As the ionized sporadic-E cloud moves, the stations you first hear fade out and are replaced by a new set of stations located more to the west and south of your location. You can often easily determine the direction in which a sporadic-E cloud is moving by the stations that fade in and out on a channel.

Sporadic-E reception time can range from just a few minutes to almost an entire day, with most sporadic-E events lasting from one to two hours. Sporadic-E clouds can form and dissipate with great rapidity; you might think a sporadic-E event is over but another cloud can form within minutes and bring in more distant stations. Signals are usually very strong during "E-skip" but are often subject to distortion and rapid fading. Sporadic-E signals can seem to suddenly appear out of nowhere, be audible for several minutes at levels exceeding those of your local stations, and then fade away to nothingness. Because signals are so strong, you don't need a fancy scanner or outdoor antenna to hear plenty of sporadic-E signals.

The effects of sporadic-E are most common on the VHF Low band. Sporadic-E propagation is much more rare on the VHF High band; some studies suggest that "E-skip" is more than 30 times more common on the VHF Low band than on the VHF High band. To date, there have been no reliable reports on sporadic-E propagation on UHF.

Sporadic-E propagation will likely affect your local stations on the VHF Low band. During a sporadic-E event, your local police or fire department might find itself experiencing inter-ference from stations hundreds of miles away. Sometimes you will hear arguments between stations in different parts of the country, each accusing the other of trying to use "their" channel. In fact, sporadic-E propagation was one of the main factors in the move of many public safety agencies to UHF.

"F" Layer Ionospheric Propagation

The highest layer of the ionosphere is the "F" layer, which is found from approximately 90 to over 400 miles above the Earth's surface. The F layer is where most shortwave (below 30 MHz) signals are reflected back to Earth, and, because of the greater elevation of the F layer, the distances covered by signals reflected from the F layer are greater than those from the E layer. For example, the minimum distance covered by reflection off the F layer is about 2000 miles and can easily be 10,000 miles or more. Normally, signals above 30 MHz zip through the F layer just as easily as they do through the E layer. However, there are certain years at the peak of the sun's sunspot cycle when the F layer becomes heavily enough ionized to reflect signals in the VHF Low band as if they were shortwave signals. When this happens, your scanner can come alive with signals from Europe, Africa, Australia, Japan, China, and other spots around the globe.

The number of sunspots is the key indicator of how "active" the sun is. A large number of sunspots indicates the sun is emitting high levels of ultraviolet radiation and a large number of charged particles, both of which increase the level of ionization in the F layer of the ionosphere. During years of high sunspot activity, the highest frequency that the F layer can reflect back to Earth can rise as high as 50 MHz. (During years of low sunspot activity, the maximum frequency can be as low as 20 MHz.) The number of sunspots go through cycles lasting approximately 11 years in length. During a period of low sunspot numbers, such as 1984 to 1986, propagation by the F layer above 30 MHz is impossible. During years of high sunspot numbers, as was the case in 1989 to 1991, the 30 to 50 MHz range can produce F layer propagation for several days at a time. At the time this book was published in 1993, the sunspot cycle was declining toward a minimum that will probably be reached sometime in 1995 or 1996. After that, the number of sunspots will begin to increase and F layer propagation will be possible again by 1998 or 1999.

Unlike sporadic-E, F layer propagation has never been reported on the VHF High band and is probably impossible on those frequencies. F layer propagation is first noticed at the lower end of the VHF Low band (around 30 MHz) and often will not extend all the way through to 50 MHz. For example, the maximum frequency the F layer can reflect back to Earth might reach only to 38 MHz on a given day. On your scanner, you would find the 30 to 38 MHz range filled with stations from thousands of miles away, but above 38 MHz you would hear nothing but local stations.

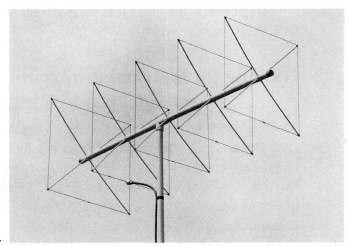

Figure 3-3: To get the most out of F layer propagation, a beam antenna such as this is very useful.

When F layer propagation is possible again starting in 1998 or 1999, the best time to listen for it will be from early October to mid-April. During the summer months, heating of the ionosphere causes it to expand and spread the ionization too thin to reflect signals above 30 MHz. F layer propagation is almost always found during the daytime. In the morning, you will hear stations to the east of your location, such as from Europe and Africa. During the middle of the day, you can mainly hear stations from Central and South America. And by late afternoon, stations from west of your location, such as Hawaii, the Pacific, Australia, and Asia will start to come in. This will continue until shortly after your local sunset, at which time the signals will fade out until after sunrise the next day.

F layer propagation seems so fantastic that many scanner fans who get interested in monitoring during a period of low sunspot activity have a hard time believing it is really possible. However, during the 1989 to 1991 period listeners in North America were able to hear stations from such countries as South Africa, Russia, Australia, and Indonesia on their scanners. Several listeners were even able to hear combat field communications during 1991's Persian Gulf War, including American tanks operating inside Iraq! This kind of reception will return before too long.

Looking for Unusual Propagation

One sure sign that some form of unusual propagation is taking place is when your local stations experience interference from other stations, particularly if police, fire, and other emergency channels are involved. Such frequencies are allocated by the Federal Communications Commission with the intent to keep interference between stations to an absolute minimum.

If your scanner covers the 50 to 54 MHz and 144 to 148 MHz ham bands, those frequencies can be a great way to keep up with unusual propagation. Hams like nothing better than to be able to talk to someone hundreds or thousands of miles away, and actively try to do so whenever possible. Programming some of the channels hams use for simplex communications (see Chapter Six) into your scanner will give you early warning that propagation is favorable for long distance reception at VHF and UHF. In particular, the 29.60 MHz simplex frequency used by hams gets very busy when F layer propagation reaches that frequency, and high activity on that channel is a good indicator that F layer propagation is possible above 30 MHz.

Some scanner listeners dedicate one of their memory banks to channels that are normally vacant in their area, dividing the available channels between VHF Low (for F layer and sporadic-E propagation) and VHF High (for tropospheric bending). During times when unusual propagation is most likely (such as autumn for "tropo" and June and July for sporadic-E) this bank is scanned whenever the scanner is in use. Other listeners—more interested in receiving distant stations than their locals—dedicate a scanner to look for distant stations and continuously use the search function of their scanners to roam vacant channels in search of unusual receptions.

The decision back in the 1940s to put local radio services on frequencies above 30 MHz was based upon the assumption that VHF and UHF were immune from the propagational effects that made long distance reception possible below 30 MHz. While it's true that VHF and UHF signals are less likely to be heard hundreds or thousands of miles away than those below 30 MHz, we now know the original assumption was very wrong. If you ever notice some sort of "funny signal" on your scanner, it could be a clue that a signal from the other side of the country or world is trying to come in.

CHAPTER FOUR

A Guided Tour of the VHF/UHF Bands

STATIONS ARE NOT SCATTERED randomly
across the scanner bands, although it can seem that
way at first. Instead, certain ranges have been set aside
for specific purposes and activities. In this chapter, we'll go
through the VHF/UHF spectrum and see how it has been
allocated among its various users.

Although we have earlier defined VHF as beginning at
30 MHz, we'll start at 25 MHz. This is because an increasing
number of scanners start their coverage at 25 MHz instead of
30 MHz. Propagationally, there isn't much difference between
25 and 30 MHz. Our tour will end at 940 MHz, which is the
upper limit of a range set aside for business communications
using trunked repeater systems. There are plenty of other users
(especially satellites) of frequencies above 940 MHz, but most
scanners do not cover this range and specialized antenna systems are necessary to get best reception.

So, if you're ready, let's see what lies above 25 MHz. . . .

25 to 25.67 MHz: This range may seem completely dead
during years of low sunspot activity, but when sunspots are
numerous it is used by military and navy stations around the
world. Most transmissions will be in SSB or even Morse code
and radioteletype. In the United States, there are a handful of
stations licensed to petroleum exploration and refining companies active from 25.02 to 25.32 MHz in FM. Channels there
are spaced 20 kHz apart (25.02, 25.04, 25.06, etc.).

25.67 to 26.1 MHz: This is another band that is quiet until years of high sunspot activity. It is allocated to international shortwave broadcasting, and was used by such countries as France, South Africa, and England during the 1989–1991 peak of the last sunspot cycle. Such broadcasts will be in AM. You will also hear scattered two-way radio communications in various languages using AM, FM, and SSB in this range. Most will be illegal "bootleg" or "outlaw" radio systems using modified citizens band (CB) radio units; sometimes these stations will sound much like legitimate ham radio stations, even down to official-sounding call signs. Finally, you may hear what sound like relays of AM and FM broadcast stations, complete with commercials and local newscasts. These are remote broadcast relay systems used to link a remote broadcast site (like a shopping mall or sporting event) with the main broadcast studio or to cue remote facilities (like traffic helicopters) that they will soon be on the air. These signals are in FM, and are found on channels spaced 40 kHz apart beginning at 25.87 and running to 26.47 MHz. However, many radio stations are moving their remote broadcast facilities to UHF and activity on these channels is dropping off.

26.1 to 26.965: This band is occupied by a hodge-podge of often unlicensed and illegal two-way radio systems from around the world. This part of the radio spectrum is a "no man's land" that seems like a radio version of the Wild West; you'll hear several different languages and modes used here. Among the illegal users of this band are fishing boats, "gypsy" taxicabs in major cities like New York and Los Angeles, truckers, and even drug dealers. Some illegal networks even include computer-to-computer linkups and on-the-air computer bulletin boards. Your best chance of hearing such activity is during years of high sunspot numbers or during sporadic-E propagation, particularly when signals are being received from Central and South America.

26.965 to 27.405 MHz: This is where you'll find the 40 channels used by CB radio in the United States and Canada. AM is mainly used up to about 27.355 MHz and SSB thereafter.

27.405 to 28 MHz: Business users are found from 27.41 to 27.53 MHz in FM using channels spaced 20 kHz apart. Several agencies of the United States government use 27.575 and 27.585 MHz for low-powered "walkie-talkies" in both AM and FM. Those same channels are also used for communications between different agencies of the U.S. government for mobile and portable communications, especially during joint operations or emergencies. A very active frequency is 27.575 MHz, used for FM communications by the Department of State, Department of the Treasury, Department of Transportation, and the Department of Health and Human Services. The Department of the Treasury also uses 27.585. The Federal Aviation Administration can be heard on 27.625 MHz in FM. But the busiest users of this frequency range are the so-called "freebanders" or "outbanders," usually in SSB. These stations are unlicensed and illegal, sounding much like ham radio stations. They are especially active during sporadic-E or F layer propagation, contacting each other from around the world.

28 to 29.7 MHz: This is a ham radio band, known as the "10-meter" band after its equivalent wavelength. From 28 to 29 MHz, most operation is in SSB with a smattering of Morse code and AM. Above 29 MHz, FM is used. 29.6 MHz is a very popular FM simplex frequency for hams and a good indicator of when F layer and sporadic-E propagation is possible above 30 MHz. Hams also operate several repeater systems around the country whose output frequencies include 29.62, 29.64, 29.66, and 29.68 MHz.

29.7 to 30 MHz: Like the 27.405 to 28 MHz range, this slice of the radio spectrum is home to both legal and illegal communications. In the United States, the forest products radio service has an allocation from 29.71 to 29.79 MHz using FM and 20 kHz channel spacing. These stations are operated by logging companies, sawmills, trucks, cutting crews, and others involved in planting and harvesting timber. There are also some military operations of the United States and other nations found here in SSB, FM, and AM. Several Canadian businesses are authorized to use this band with FM. This range is also populated with numerous illegal and bootleg communications networks.

30 to 50 MHz: This is a workhorse band for many government services and businesses, with most operating on channels spaced 20 kHz apart. While there are numerous exceptions and local variations, here is the general usage plan for this range:

30.01–30.56	U.S. government
30.66–31.24	Industry
	Forestry and logging
32.26–31.98	Emergency services
	Business
32.00–33.00	U.S. government
33.02–33.16	Highway maintenance
	Emergency services
	Business
33.18–33.38	Petroleum operations
33.42–33.98	Fire departments
34.00–35.00	U.S. government
35.02–35.18	Business
35.22–35.66	Mobile telephones and paging
35.70–35.72	Business
35.74–35.98	Emergency services
	Business
36.00–37.00	U.S. government
37.02–37.44	Local government
	Law enforcement
37.46–37.86	Utility companies
37.90–37.98	Highway maintenance
	Emergency services

38.00–39.00	U.S. government
39.02–39.98	Local government
	Law enforcement
40.00–42.00	U.S. government
42.02–42.94	State police
42.96–43.18	Special industrial
	Business
43.22–43.68	Mobile telephones and paging
43.70–44.60	Trucks
	Business
44.62–45.06	State police
	Forestry
	Conservation
45.08–45.66	Law enforcement
45.68–46.04	Law enforcement
	Emergency services
	Highway maintenance
46.06–46.50	Fire departments
46.52–46.58	Local government
46.60–47.00	U.S. government
47.02–47.40	State government
	Highway maintenance
47.42	Red Cross
47.44–47.68	Special industrial
	Emergency services
47.70–48.54	Utility companies
48.56–49.58	Forestry
	Petroleum operations
49.60–50.00	U.S. government

In addition to the users above, low power "walkie-talkie" units are authorized to operate around 49 MHz. These are the type popular as kids' toys, and all transmissions are in FM. Cordless telephones also operate in this frequency range. Base station cordless telephones operate from 46.61 to 46.97 MHz while the remote portable handsets operate from 49.67 to 49.97 MHz.

50 to 54 MHz: This is another ham radio band, known as "6-meters." The first 600 kHz is mainly SSB, with a little AM. The rest of the band is mostly FM, both simplex and by repeaters. 52.525 MHz is the most popular FM simplex channel, and is a good indicator of sporadic-E and F layer propagation. You may hear some unusual tones above 53 MHz; these are hams transmitting remote control signals.

54 to 72 MHz: This is where television channels 2, 3, and 4 are located. You'll be able to hear the audio portion of TV signals on your scanner, although they will be usually be distorted unless your scanner has a wideband FM mode. The video portion will sound like a loud "buzz."

72 to 76 MHz: This range is mainly used for in-factory communications by manufacturers and remote control of model airplanes and other devices, such as garage door openers. Most channels here are spaced 20 kHz apart. Since many scanners do not cover this range, it is also popular with law enforcement and private investigators for "body microphones," tracking transmitters, and other devices for remote surveillance and eavesdropping. Many of the roadside call-boxes used to summon assistance also operate here. Most channels are spaced 20 kHz apart.

76 to 88 MHz: This is where television channels 5 and 6 are found.

88 to 108 MHz: This is the FM broadcasting band. Signals will be distorted unless your scanner can receive wideband FM.

108 to 136 MHz: Civilian aviation uses this band. The 108 to 118 MHz range is used for air navigation beacons operating under the

Figure 4-1: The IC-R1 by Icom is a high-performance hand-held scanner with 100 memory channels.

"Omni" system. This is a range you might want to avoid scanning or searching, since beacons operate continuously and will cause your scanner to "freeze" on their channels. Like all transmissions in this band, beacons use AM and transmit such material as their identification in Morse code, "beeping" sounds, and aviation weather bulletins. The 118 to 136 MHz segment is far more interesting; here you can monitor air traffic control centers, airport control towers, planes arriving and departing from an airport, and airport operations. Most major airports use several different channels. For example, one channel might be for arriving flights, another for departing flights, and a third could be used by the control tower for overall operations. Most channels are spaced 25 kHz apart.

136 to 138 MHz: Weather satellites use this range to transmit photographic weather maps continuously. The United States, Russia, and Japan all have satellites operating here. Some people have installed the necessary receiving equipment and receive the maps directly in their homes. If your scanner can tune this range and you have a good outdoor antenna, you may be able to hear signals from the U.S. "Tiros" satellites on 137.15 and 137.30 MHz and the Russian "Meteor" satellites on 137.50 and 137.62 MHz.

138 to 144 MHz: In the United States, this band is used mainly by the various military services for their land mobile communications. Channel spacing varies across the country, although most operations take place on channels at 5 kHz intervals. These frequencies are used at various military bases for operations such as security, maintenance, traffic control, etc. 143 to 144 MHz has several stations belonging to the military affiliate radio service (MARS). These stations are operated by ham radio operators under the auspices of various military services. These hams assist the military with non-critical tasks such as relaying messages from service personnel abroad to their families back home.

144 to 148 MHz: This is the most popular ham radio band, known as "2-meters." The first 400 kHz is used for mainly for SSB, and the rest of the band is FM. There are several hundreds of repeater stations for hams across the country on this band, and there are likely several you can hear on your scanner. Chapter Six describes VHF and UHF ham radio in more detail, but some popular repeater output frequencies include 146.76, 146.82, 146.88, and 146.94 MHz. 146.52 MHz is the most popular simplex frequency. Spacing between channels in this band varies from 10 to 15 to 20 kHz, with almost all operations above 144.50 MHz taking place at 5 kHz intervals.

148 to 150.8 MHz: This band is similar to 138 to 144 MHz, with several MARS stations found on 148.01 MHz. The civil air patrol (CAP) uses 148.150. Various military services use 148.155 to 148.25 MHz, while the U.S. Navy uses 148.29 to 150.75 MHz exclusively.

150.8 to 174 MHz: The VHF High band is another busy place. Most channels are spaced 15 kHz apart, although this may vary slightly in different areas. Here is the general usage pattern you'll find:

150.815–150.995	Business
151.01–151.30	Highway maintenance
151.145–151.475	Forestry
151.505–151.595	Special industrial
151.625–151.955	Business
151.985–152.24	Mobile telephone
152.27–152.45	Taxicabs
152.48–152.84	Mobile telephone and paging
152.87–153.02	Special industrial
	Motion picture production
153.05–153.44	Petroleum operations
	Forestry
153.47–153.71	Power utilities

153.74–154.115	Local government
154.13–154.445	Fire departments
154.45–154.60	Business
	Petroleum operations
	Special industrial
154.655–155.145	Local government
	Law enforcement
	State police
155.16–155.40	Emergency services
	Law enforcement
155.415–156.03	Local government
	Law enforcement
156.045–156.24	Highway maintenance
	Law enforcement
156.275–157.425	Marine (boats and coastal stations)
157.455–157.50	Automobile emergency
157.53–157.71	Taxicabs
157.74–158.10	Mobile telephone and paging
158.13–158.46	Forestry
	Petroleum operations
	Utility companies
158.49–158.70	Mobile telephone
158.73–158.97	Law enforcement
	Local government
158.985–159.21	Law enforcement
	Highway maintenance
159.225–159.465	Forestry
	Conservation
159.51–160.20	Trucking
160.215–161.565	Railroads
161.60–162.00	Marine (boats and coastal stations)
162.025–162.175	U.S. government
162.40	Weather broadcasts
162.55	Weather broadcasts
162.75–163.25	U.S. government

163.275	Weather broadcasts
163.30–163.535	U.S. military
163.50–169.30	U.S. government
169.45–169.725	Special industrial
170.15–170.22	U.S. government
170.225–170.325	Land transportation
170.425–170.575	Forestry
170.975–171.25	U.S. government
	Land transportation
171.385–172.775	U.S. government
173.025	Weather broadcasts
173.075	U.S. government
173.1–173.975	Land transportation
	Petroleum operation
	Press services

A new and interesting frequency is 151.625, which is an itinerant operations channel used largely by hand-held portable units. Traveling operations, such as circuses, balloons, and search and rescue units, use this channel. Other users include several major league sports teams. The 162 to 174 MHz range is extensively used by U.S. government agencies.

174 to 216 MHz: Television channels 7 through 13 are located here.

216 to 220 MHz: The automated maritime telecommunications system (AMTS) operates in this range. This is a communications system used by boats and shore stations on major inland waterways (such as the Great Lakes and the Mississippi and Ohio rivers) and the Gulf of Mexico. Channels are spaced 12.5 kHz apart and all transmissions are in FM.

220 to 222 MHz: Until recently, this range was allocated to ham radio operators. It was recently re-assigned to the land mobile service. The channel spacing and usage patterns had not been finally decided at the time this book was written.

222 to 225 MHz: This is another ham radio band, similar in many respects to the 144 to 148 MHz ham band. Channel spacing is generally in 20 kHz increments. There are numerous repeaters here, with output frequencies running from 223.92 to 224.98 MHz. The most popular simplex channel is 223.50 MHz.

225 to 400 MHz: This huge allocation is devoted exclusively to military aviation. Like the civilian aviation band, all transmissions will be in AM. Channels are usually spaced at 100 kHz intervals, although some operations take place at 50 kHz intervals. Among the more active channels nationally are 311 and 319.4 MHz by the U.S. Air Force, 282.8, 317.7, and 317.8 MHz by the U.S. Coast Guard, and 340.2 by the U.S. Navy. The Space Shuttle can be heard on 259.7 and 296.8 MHz during its landing phase of each flight; if you live within about 200 miles of the Shuttle's California and Florida landing sites, these channels are worth monitoring.

400 to 406 MHz: This is something of a hodge-podge occupied by an assortment of military and government stations, most in FM.

406 to 420 MHz: This is an exclusive U.S. government band, with most stations using FM on channels spaced 25 kHz apart. Air Force One and other aircraft used by the executive branch of the federal government can be heard on 415.7 MHz, usually carrying phone calls from the aircraft. The U.S. Postal Service can be heard on 413.60, 414.75, 415.05, 416.225, and 418.30 MHz.

420 to 450 MHz: This is a ham radio band, second only to the 144-148 MHz range in popularity above 50 MHz. The 420 to 444 MHz segment is used for SSB, ham television (that's right— hams are authorized to use television on this band), and ham radio communications satellites. Repeater stations are found from 444 to 450 MHz.

450 to 470 MHz: This band has grown in popularity among land mobile radio systems trying to escape the disruptive effects of sporadic-E and F layer propagation on the VHF Low band. While there are numerous exceptions and local variations, here is the general usage plan for this range:

450.05–450.95	Remote broadcast relays
451.00–451.150	Utility companies
451.175–451.75	Forestry
	Petroleum operations
	Telephone maintenance
451.775–451.975	Special industrial
452.00–452.50	Railroads
	Taxicabs
452.525–452.60	Automobile emergency
452.625–452.975	Railroads
	Trucking
453.00–453.975	Fire departments
	Law enforcement
	Local government
454.00–454.975	Mobile telephone
455.00–455.975	Remote broadcast relays
456.00–458.975	Fire departments
	Land transportation
	Law enforcement
459.00–459.975	Business
	Special industrial
460.00–460.625	Fire departments
	Law enforcement
460.65–462.175	Business
462.20–462.45	Taxicabs
462.55–462.725	General mobile radio service (GMRS)
462.75–462.95	Business
463.00–463.175	Medical
463.20–464.975	Business

465.00–467.50	Fire departments
	Industrial
	Land transportation
	Law enforcement
467.55–467.725	General mobile radio service (GMRS)
467.75–467.925	Business
467.7375–469.975	Industrial
	Land transportation

470 to 512 MHz: This is the UHF-T band, and shares the same frequencies used by television channels 14 to 20. Usage of this band varies across the country depending on whether there are any local television stations using channels 14 through 20. Several big city police departments have moved their operations up to this band.

512 to 825 MHz: This range is used by television channels 21 through 72.

825 to 849 MHz: Mobile cellular telephone units use this band to transmit to cells. The frequencies used by a moving cellular telephone will abruptly switch depending upon which cell is handling a call.

849 to 851 MHz: This is the range used to provide telephone service from airplanes. Unlike other mobile telephone services, a form of SSB is used here and you will need a scanner capable of receiving SSB.

851 to 866 MHz: This range is mainly used by businesses and a handful of public safety services. Channels are spaced 25 kHz apart.

Figure 4-2: The GMR100 by Uniden is a portable unit designed for use in the general mobile radio service.

866 to 869 MHz: Public safety services such as fire departments and law enforcement are the major users of this segment.

869 to 894 MHz: This range is used by the cells in a cellular telephone system to transmit to mobile telephones.

902 to 928 MHz: This is a relatively new ham radio band which is lightly used at present. It is shared with land mobile communications. In most parts of the country this will be a quiet segment of the radio spectrum. However, the FCC has announced that this range will be used by new generation cordless telephones and other personal communications devices such as remote control units and room monitoring devices. Some of these will use "spread spectrum" technology that will make eavesdropping on such signals almost impossible.

928 to 932 MHz: This is used by radio paging systems. Channels are spaced 25 kHz apart.

935 to 940 MHz: This is a new allocation for business radio systems using trunked repeaters. Channels are spaced 12.5 kHz apart.

To get a good idea of specific channels in use in your area and what you can hear, consult the publications listed in Appendix A. If there is a scanner club active in your area, as listed in Appendix A, then you should join it to keep up with latest channel changes and new users.

CHAPTER FIVE

Land Mobile Radio Services

MOST OF WHAT YOU HEAR on your scanner will be communications from stations in one of the land mobile radio services. A land mobile system normally consists of a base station, with coverage possibly augmented by a repeater system, and one or more mobile stations. However, it's also possible for some land mobile systems to consist of all mobile units, such as the systems used for communications at circuses, golf tournaments, and other events which are held on an irregular basis or that travel around the country.

For specific frequencies in use in your area, particularly for police and fire departments, you need to consult one of the publications listed in Appendix A. A good way to keep up with changes in your area is to join a scanner club; a list of these is also given in Appendix A. In this chapter, we'll look at some frequencies used nationally and the general operating patterns in the land mobile services.

State and Local Government

There are some frequencies widely used by state and local governments throughout the United States. One is 154.28 MHz, which is used for mutual aid communications by many fire departments. If a fire is so severe that assistance is requested from neighboring communities, this channel will likely be active. 155.475 MHz is used for similar mutual aid communications by many police departments. There has been a steady trend for local police and fire services to move up to the 800

MHz band and use trunked repeaters for extended coverage. In fact, the bulk of police and fire communications in the nation's 20 largest metropolitan areas are either already operating on 800 MHz or are in the process of moving there. Mutual aid frequencies on 800 MHz include 866.0125, 866.5125, 867.0125, 867.5125, and 868.0125 MHz. Frequency ranges for police, fire, and other local emergency services can be found in Chapter Four.

Many state police and highway patrol departments still use frequencies in the VHF Low band for maximum coverage. These can be found between 42.02 to 42.94 MHz. Both local and state law enforcement agencies make extensive use of the "10-code," which is a sort of radio shorthand involving the using of numeric codes beginning with 10. An extensive 10-code listing is in Appendix B.

Many state agencies use the VHF Low band for their operations. Typical users include state park personnel, public works departments (including highway maintenance), and conservation agencies.

Federal Government

The civilian sectors of the United States government have increasingly abandoned the VHF Low band in recent years, but there is still some activity there by agencies whose operations often involve large coverage areas. An example is the U.S. Fish and Wildlife Service, which can be heard on 34.25, 34.41, 34.81, 34.83, and 34.85 MHz.

The federal government is an active user of the VHF High and UHF bands on a multitude of frequencies. Fortunately, some channels are used throughout the country. One hot frequency is 166.4625 MHz, used by the Department of the Treasury and its agencies, including Customs, Secret Service, and Bureau of Alcohol, Tobacco, and Firearms. All two-way radio equipment used by the Treasury agencies will have this frequency available, and it's often used as a common communications channel for joint operations by those agencies. Similarly, the Federal Bureau of Investigation uses 167.5625 MHz as its national communications channel. Local offices often use different frequencies for their daily communications, but every FBI radio can operate on this frequency. The Secret Service also makes heavy use of 165.375 MHz nationally.

Some less glamorous federal agencies can be heard as well. The Interstate Commerce Commission uses 409.20 MHz nationally, and the Federal Communications Commission does the same with 167.05 MHz. The Department of Housing and Urban Development uses 164.50 MHz in many parts of the country. The Department of State's common frequency is 409.625 MHz. You don't have to be near Washington, DC, to hear these and other stations. For example, the State Department uses 409.625 MHz in cities like Dallas, Miami, Los Angeles, and even Oak Ridge, TN.

Canadian listeners are not left out. Agriculture Canada uses 464.1875 MHz, Customs and Immigration is on 141.27 and 406.025 MHz, and the National Film Board of Canada uses 171.87 MHz.

Businesses and Industries

In the United States, there are several radio services open to businesses and other organizations (schools, hospitals, etc.) that require two-way communications. The Business Radio Service is open to virtually any business that needs two-way radio for its operations, and there are several specialized services, like the Taxicab Radio Service, available for certain types of businesses. The Special Industrial Radio Service includes construction crews, surveyors, crop dusters, shipyards, mining, ranching, and many other industries. Frequency ranges for special industrial communications were given in Chapter Four. Forestry and petroleum operations (oil exploration, extraction, and refining) have several segments for their use in the VHF Low and High bands, also detailed in Chapter Four. The Manufacturers' Radio Service has two allocations in the 72-76 MHz band, 72.02 to 72.6 and 75.44 to 75.6 MHz, as well as 153.05 to 153.395, 158.28 to 158.43, 451.175 to 451.675, and 462.20 to 462.525 MHz.

In recent years, several fast-food restaurant chains have been using two-way radio to send customer orders from the drive-through window to the kitchen. You've probably seen the headsets used with the radio sets worn by personnel manning the drive-through window. McDonalds restaurants have been heard on 151.895, 154.60, 170.245, and 171.105 MHz. The Burger King chain has been heard on 465.887 and 467.787 MHz, while several Taco Bell restaurants use 465.887 MHz. You might think that listening to someone relay customer orders to the kitchen would be dull listening, but the remarks about customers and the food they order can be hilarious!

Disaster Communications

During local and regional emergencies, you may be able to hear Red Cross relief efforts on 47.42, 47.46, 47.50, 47.60, and 47.66 MHz. If a major disaster, such as an earthquake or hurricane, has happened near you, you can listen to Federal Emergency Management Agency (FEMA) operations on 142.23, 142.35, 142.425, 142.975, 143.00, 164.8625, and 165.6625 MHz. If the National Guard has been called out, two common frequencies for them are 34.90 and 163.4875 MHz, although these may vary in some states.

If a situation exists where civil authority breaks down (as was the case with the 1992 Los Angeles riots), the U.S. Army and Marine Corps may be called out to restore order. The national military disaster preparedness frequency is 163.5125

MHz. Among the frequencies observed in use by the U.S. Army during the Los Angeles riots were 141.06, 141.12, 141.465, and 142.44 MHz.

Itinerant Operations

Some channels in the business and special industrial band allocations have been set aside for use by itinerant users. As defined by the Federal Communications Commission, an itinerant user is one that operates a two-way communications system at different unspecified locations for varying lengths of time. As a practical matter, any system that uses hand-held portable units exclusively qualifies for operation on itinerant channels, even if all hand-held units are used at a single location.

You don't have to be a business or industrial corporation to get a license to operate on the itinerant frequencies; the FCC will issue a license to virtually anyone. As a result, itinerant channels are used by farmers, hunters, hikers, campers, and just about any other category of people who need short range communications. The growing availability of transceivers for itinerant use has resulted in their use by many small companies and individuals who have never bothered to get a license. As a result, you'll hear everything from the family down the street to gangs and drug dealers on itinerant channels.

Most "walkie-talkies" used on itinerant channels are low powered, usually only a watt or two. This gives them a maximum mobile-to-mobile range of less than five miles, depending on terrain. More typical range is about two miles or so, so most of the communications you hear will be near you. You may hear some higher powered base stations on itinerant channels, but these base stations are permitted by the FCC to be used at different locations on a temporary basis. (Base stations in other services can only operate from the location specified on the station license.)

Figure 5-1: This transceiver from Radio Shack comes equipped for operation on the 151.625 MHz itinerant channel.

The most widely used itinerant channel is 151.625 MHz, with 154.57 and 154.60 MHz also being popular. Itinerant stations on the VHF High band can show up anywhere from 154.655 to 151.955 MHz and again from 154.515 to 154.54 MHz. On the VHF Low band, 35.04 MHz is heavily used. UHF channels include 464.50, 464.55, 469.50, ad 469.55 MHz. In Canada, 167.73 MHz is the most popular itinerant channel. Other Canadian channels include 32.48, 32.52, 32.56, 462.50, and 462.90 MHz.

Finally, the UHF-T band is often used for itinerant operations during really major events like a Super Bowl, Olympics, or political convention. The size and scope of such events can overload the normally allocated itinerant channels, so the FCC often temporarily assigns channels in the UHF-T band for the duration of the event. If such a major event is taking place in your area, the UHF-T band could be a great place to look for "behind the scenes" communications.

General Mobile Radio Service (GMRS)

The general mobile radio service (GMRS) was actually the very first citizens band (CB) radio service. It was established soon after World War II and was known for years as the Class A CB service (the CB radio band near 27 MHz was known as the Class D CB band). For years, the Class A band languished because equipment for UHF was expensive and coverage in pre-repeater days was limited. By the time equipment prices became more reasonable and repeater technology was available, the Class D CB boom was underway and Class A was forgotten.

In the late 1970s, the FCC made several moves to revitalize Class A. The transmission mode was switched from AM to FM, repeater stations were authorized, and the name was changed from Class A CB to GMRS to make it clear this was a service distinct from Class D CB.

GMRS can be thought of as a business radio service open to everyone. Unlike CB, you can't pick any channel you want and start operating. There are eight main GMRS channels, and the GMRS license from the FCC will generally restrict you to operation on a specified channel. These channels are paired with eight additional frequencies for repeater inputs. Mobile and hand-held units can transmit on the repeater input and be relayed on the repeater output/base station frequency. Mobile and portable units are also able to use simplex on the repeater/base station frequency.

There are some exceptions to the single channel restriction in GMRS, however. 462.675 MHz is used throughout the United States for emergency communications and traveler assistance. GMRS licensees can operate on 462.675 MHz for those purposes regardless of what their assigned channel is, and there are numerous repeaters that operate on the 467.675/462.675 input and output frequency pairing. In addition, there are six "splinter" channels, as shown in Table 5-1, available for simplex operation. Any GMRS licensee can use these channels in addition to their primary specified channel. Most operations on the splinter channels are conducted with low powered hand-held units.

Many GMRS users on a channel will share a repeater and employ some type of coded access to minimize unwanted calls. GMRS channels have become a favorite of nongovernmental, volunteer assistance organizations such as mountain search and rescue teams, ski patrols, rescue squads, etc. GMRS channels are also popular with boating, hiking, and other sporting clubs and even families that need to reliably stay in touch with each other over a wide area.

Table 5-1

General Mobile Radio Service Frequencies

Mobiles/Repeater Input	Base/Repeater Output	"Splinter" Channels for Simplex
467.55	462.55	462.5625
467.575	462.575	462.5875
467.60	462.60	462.6125
467.625	462.625	462.6375
467.65	462.65	462.6625
467.675	462.675	462.6875
467.70	462.70	
467.725	462.725	

Movies and the Press

Filming a motion picture on an elaborate set, particularly an outdoor one, requires good communications between the director and crew members. The Federal Communications Commission has established a special motion picture radio service for that purpose. If a movie is being shot in your area, try searching the 152.87 to 153.02 and 173.225 to 173.375 MHz ranges for communications relating to the filming. Monitors in southern California and New York sometimes get to hear some pretty candid evaluations of various stars by the crew that works with them!

Newspapers also have their own frequency segments. If you're lucky, you might get to hear reporters and photographers being dispatched to the scene of major news events and, later, their eyewitness accounts. More typically, most of the communications will take place early in the morning and will

Figure 5-2: This transceiver from Maxon is typical of the units designed for operation on the new GMRS "splinter" channels.

involve delivery of papers. Try searching 173.225 to 173.375 and 452.975 to 453.00 MHz for stations in your area.

Broadcast relay stations are used when radio stations cover events remote from the studio location, and include everything from football games to grand openings of shopping centers. These comments made by the announcers during station breaks and commercials are a lot of fun to hear because most announcers don't realize there are still some people who might hear them. Search the 450.05 to 450.95 and 455.00 to 455.975 MHz segments for them.

Trains, Trucks, Cabs, and Buses

A lot of railroad fans use their scanners to "ride the rails." The 160.215 to 161.565 and 452.325 to 452.95 MHz ranges are especially busy with communications about the coupling, decoupling, switching, and dispatching of trains and their cars. These channels are also used for communications between the locomotive and caboose of a train.

It's easy to hear other land-based transportation systems. Buses and trucks can be found in the 30.66 to 31.14, 43.70 to 44.60, 159.495 to 160.20, and 452.325 to 452.875 MHz ranges. Taxicabs may use GMRS or other business channels, but they are most common in the 152.27 to 152.465 and 452.05 to 452.50 MHz ranges for base stations and 157.53 to 157.725 and 457.05 to 457.50 MHz ranges for the cabs themselves. What if one of the taxis breaks down? Listen for towing trucks in the 150.815 to 150.965, 157.47 to 157.515, and 452.525 to 452.60 MHz ranges.

As we'll see in Chapter Eight, many large cities have fleets of "gypsy" cabs that generally do not have the necessary city licenses and operate in violation of city laws. Not surprisingly, these cabs use two-way radio equipment without the benefit of FCC licenses and on frequencies not allocated to taxis or other motor transport.

CHAPTER SIX

Aeronautical, Marine, and Ham Radio Communications

VIRTUALLY ALL SCANNERS TODAY will let you listen to marine communications on the 156.275 to 157.425 and 161.60 to 162.00 MHz ranges of the VHF High band. These frequencies are used by the U.S. Coast Guard, commercial vessels, and weekend boaters for communications with marinas, the Coast Guard, and between vessels. This band is in use along ocean coasts as well as inland waterways. An increasing number of scanners are capable of tuning the 108 to 136 MHz civilian aeronautical band and the sometimes dramatic activity there. Finally, ham radio operators have several allocations at VHF and UHF, including the very popular 2-meter (144 to 148 MHz) band. If you decide you'd like to talk to the ham radio operators you hear on your scanner, it's easy to get your own ham license and join the fun. Let's take a closer look at these specialized ranges.

Civil Aeronautical

One big difference in this band is that all communications are in AM, not FM. Scanners covering the aeronautical band will typically switch automatically to AM when tuning it. The only difference you'll notice is that noise will probably be greater than on the other bands that use FM, particularly "crashes" produced by lightning. This is because AM has lower immunity to noise than FM.

Another big difference is the range covered by aeronautical communications. Ground stations, such as airport control towers, will have a coverage area roughly equal to stations in the VHF High band. However, you will be able to hear airplanes aloft at surprising distances from you, often over 300 miles away. As you probably suspect, this is because the horizon, and resulting line-of-sight range, is much greater at 30,000 feet than at 300 feet. You will be able to hear airplanes communicating with several ground stations that you are unable to hear from your location. By the way, aeronautical communications are generally simplex.

The 108 to 136 MHz band is used for civilian aviation only. There is an enormous band from 225 to 400 MHz used for military aviation, and that will be discussed in Chapter Seven.

The very first aeronautical frequency you should program into your scanner is 121.50 MHz, the international aeronautical emergency channel. This channel is monitored by aeronautical facilities worldwide, and can provide some dramatic listening when active. Emergencies can range from mechanical problems, to bad weather, to a mentally disturbed passenger on board. If you get seriously interested in monitoring aircraft, this should be a priority channel for your scanner.

The 108 to 118 MHz range will probably be pretty much of a loss, since this is where "Omni" system air navigation beacons operate. If you tune this range, all you will hear will be stations continuously repeating their identification in Morse code or a series of coded "beeps" along with some voice weather bulletins. Since these stations transmit continuously, your scanner will stop at the first one it finds and "freeze" on the channel. That's why trying to search 108 to 118 MHz isn't a good idea.

The 118 to 136 MHz segment is much more interesting. The communications you hear in that range involve several different types of stations. At major airports, the *control tower* is the hub of communications activity with aircraft. There will be a separate channel used by the control tower. At many airports, however, the task of controlling arriving and departing flights is not done by the control tower, or on the control tower frequency. Instead, two separate operations and frequencies for each are used. When a commercial flight is preparing to leave an airport, you will hear the airport's *departure control* directing the aircraft. When a commercial flight is preparing to land at an airport, *approach control* directs it in to a landing. Both departure and approach control are staffed by air traffic controllers. Once the flight is airborne and en route to its destination, control will be handed over from the airport to a regional *air traffic control center*. The air traffic control center will operate on a different channel than either approach or departure control. For example, a flight to Los Angeles may be directed by the air traffic control center at Las Vegas. As the flight approaches Los Angeles, the Las Vegas air traffic control center will transfer flight control to Los Angeles approach, and the aircraft will switch to the channel used by Los Angeles for that purpose. Once the flight lands in Los Angeles (or a similar major airport), control of the airplane's movements will be handed over to *ground control*. Ground control will direct the airplane to its gate, as well as control movement of utility vehicles on the runway, and operates on yet another separate channel. Ground control also directs fire and emergency vehicles at the airport in case of a problem or accident.

Major airports have other channels in use. One common one is the *automatic terminal identification service*, which is a continuous recorded broadcast of weather and runway conditions and a description of instrument flight facilities at the airport. This broadcast is regularly updated as conditions change. Other channels may be used for instrument landing systems or an electronic course correction system known as *Vortac*. Both of these operate in the 108 to 118 MHz segment.

There are other channels in use at many airports. One is the *common traffic advisory frequency*. This is a channel used for communication between all aircraft at a certain airport. Another is *Unicom*, which is a channel that lets pilots directly contact ground facilities. At smaller airports, the Unicom channel is used for some air traffic control functions. For small airports without a formal communications facility, a *Multicom* channel is used. The most common Multicom frequency is 122.90 MHz.

Smaller private aircraft make use of a *flight service station.* Flight service stations provide information on weather and airport conditions, flight plans, restricted flight areas, and pilot reports. Flight service stations are not usually at airports, but instead are located at various points along popular flight routes. Flight service stations use channels separate from those used by the regional and local air traffic controllers in an area.

You will also run across some *aeronautical radio service* (known as ARINC) channels in use. ARINC is used by airlines themselves for internal communications between the companies and their aircraft aloft. Among the topics you'll hear discussed are flight operations, personnel matters involving the flight crew, passenger matters (such as who needs to make which connecting flights), cargo handling, and other administrative items. These concerns are important to the airline involved, but aren't important enough to justify handling via the channels used for air traffic control. Instead, they are handled via the ARINC segment running from 128.825 to 132.00 MHz. New trunked ARINC systems are coming into use from 861.00 to 866.00 MHz. ARINC channels can make for fascinating listening; the flight crew and ground personnel can be very candid in some of their comments and opinions about the airline, its equipment, and its policies! A similar service for charter airline flights can be heard on 122.825, 122.875, and 122.950 MHz.

Airlines also use two-way radio systems at airports for baggage and cargo handling, maintenance, security, etc. However, such communications are conducted in the business segments of the VHF and UHF bands instead of the aeronautical band.

Not everything you'll hear on the aeronautical bands will be a powered, fixed-wing aircraft. National frequencies for use by gliders include 121.95, 122.75, and 123.125 MHz. Hot air balloons can be found on 122.50, 123.125, 123.40, and 464.50 MHz among other frequencies. Blimps can be heard on such frequencies as 121.92, 122.775, and 123.50 MHz.

Helicopters can also be heard on the aeronautical band. The most popular helicopter-to-ground channels are 123.05 MHz and 123.075 MHz, with 123.025 MHz used for helicopter-to-helicopter communications. Helicopters used for airborne traffic reports can often be found sending their reports on the remote broadcast segments at 450.00 to 451.00 and 455.00 to 456.00 MHz. These channels will be FM, not AM, and are used in addition to the other helicopter channels.

Table 6-1 lists some of the more active civilian aviation frequencies. For the channels used by airports and airlines in your area, refer to a VHF/UHF frequency guide for your area.

Table 6-1

Civil Aviation Channels

Frequency in MHz	Main Use
121.50	Aircraft emergency
121.60	Ground control
121.65	Ground control
121.70	Ground control
121.75	Ground control
121.80	Ground control
121.85	Ground control
121.90	Ground control
121.92	Blimps
121.95	Flight schools
	Gliders
122.00	Flight service stations
122.05	Flight service stations
122.10	Flight service stations
122.15	Flight service stations
122.20	Flight service stations
122.25	Blimps
	Hot air balloons
122.30	Flight service stations
122.35	Flight service stations
122.40	Flight service stations
122.45	Flight service stations
122.50	Flight service stations
122.60	Flight service stations
122.70	Unicom
122.725	Unicom
122.75	Air shows
	Aircraft-to-aircraft
	Gliders
	Skywriters
122.775	Blimps
	Gliders
	Hot air balloons
122.80	Unicom
122.85	Air shows
	Aircraft-to-aircraft
	Skywriters

Table 6-1 *continued*

Frequency in MHz	Main Use
122.90	Air shows
	Aircraft-to-aircraft
	Gliders
	Multicom
	Search and rescue operations
	Skywriters
122.925	Air shows
	Aircraft-to-aircraft
	Search and rescue operations
	Skywriters
122.95	Unicom
122.975	Unicom
123.00	Unicom
123.025	Helicopter-to-helicopter
123.05	Helicopter-to-ground
123.075	Helicopter-to-ground
123.10	Air shows
	Aircraft-to-aircraft
	Search and rescue operations
123.125	Air shows
	Blimps
	Gliders
	Hot air balloons
123.20	Flight schools
123.30	Air shows
	Blimps
	Flight schools
	Gliders
	Hot air balloons
123.40	Air shows
	Blimps
	Flight schools
	Hot air balloons
123.45	Aircraft-to-aircraft
123.50	Blimps
	Flight schools
	Gliders
	Hot air balloons

Marine Communications

You don't have to be located near the ocean to hear plenty of activity on the marine segments of the VHF High band. Those bands are in use on rivers, lakes, and other inland waterways as well as the high seas. Users include the U.S. Coast Guard, commercial vessels, and pleasure boaters.

Unless you're located directly on the water, you won't be able to hear too many ships. That's because marine communications at VHF are deliberately designed to be short range. Maximum coverage radius by a ship or boat on the water is only about 25 miles. Because they use better antennas and higher power, shore stations can be heard further inland.

Users of the VHF marine channels are divided into commercial and noncommercial categories. Commercial users include fishing boats, tugboats, ferries, and other vessels operated for a business purpose. Noncommercial vessels include all recreational and pleasure vessels, including everything from sailboats to private yachts. Most marine communications are via simplex, although some half-duplex channels are in use. Equipment for the marine segments of the VHF High band is similar to that for the land mobile services. In fact, some recreational boaters use nothing more than a hand-held portable unit. Repeater stations are not used. All VHF marine radio units must have a low-power setting of one watt, which permits communications within only a radius of about two miles. This allows two ships close to each other, or a ship near the shore, to communicate without causing interference to other stations on the channel. Commercial vessels are required to be equipped for VHF marine radio, while it is optional for most smaller recreational boats.

All ships that are equipped with VHF marine radio gear must have two channels. One is channel 16 (156.80 MHz), which is the distress, safety, and calling frequency. All ships are required to monitor this channel continuously unless they are in contact with another station on a different channel. Shore stations and the U.S. Coast Guard also monitor channel 16. If a ship has an emergency and requires immediate assistance, it makes its distress call on channel 16 since it is so heavily monitored. Channel 16 is where all communications relating to the safety of vessels and people are made. These can cover anything from a medical emergency involving a passenger to unexpected hazards like oil and sewage spills. Since all ships must monitor channel 16 when not actually in communications with another station, it makes an ideal "calling" channel. If a ship needs to make contact with another ship or a shore station, it can call the other station on channel 16. When the desired station responds, both stations then switch to another channel to talk. When they finish their communications, both stations return to channel 16 to monitor it for further calls. Some VHF marine radio gear makes channel 16 a priority channel, automatically switching to channel 16 if a call is received on that channel. The U.S. Coast Guard also uses channel 16 to establish contact with ships on the water. If you're curious about the VHF marine band, 156.80 MHz is where to start. If you can't hear anything on that channel, then you're unlikely to hear anything else on the VHF marine allocations.

Figure 6-1: This VHF FM marine transceiver from Midland is capable of operation on up to 120 channels. Note the large front panel button that immediately switches the unit to channel 16, the distress, safety, and calling frequency.

The other channel that all vessels must be equipped with is channel 6 (156.30 MHz). This is the intership safety channel, and is used for communications between ships to insure the safety of both vessels. For example, this channel might be used when transferring passengers or cargo between two ships on the water. Most communications here will use the one watt power level, so it's unlikely you'll hear anything here unless you're on the water and use a good outdoor antenna. A similar channel is channel 13 (156.55 MHz), which is used by the masters of commercial vessels to alert each other of their intentions and communicate as they maneuver their ships around each other. Recreational boaters often listen to channel 13 to determine where commercial vessels near them are heading.

Table 6-2 shows the most commonly used VHF marine channels. Channel 17 (156.85 MHz) is used by state or local government agencies having some jurisdiction over the water-way, such as fish and game officers or state boating authorities. Channel 9 (156.45 MHz) can be used for communications between commercial and recreational ships, while channel 1A (156.05 MHz) can be used for communications between all types of ships and shore facilities. A frequency that is "intership" can be used by ships on the water to contact each other, while a "ship-to-shore" channel can be used by a ship to contact shore facilities. Those channels allocated for international use are mainly used outside the territorial waters of the United States. The "port operations" channels are used to communicate with shore facilities serving ships. Channel 15 (156.75 MHz) is used exclusively for weather bulletins transmitted by shore stations. "Public correspondence" channels are used to place telephone calls through a shore-based operator.

Table 6-2

VHF Marine Channels

Channel	Shore	Ship	Use
1A	156.05	156.05	Port operations and intership
5A	156.25	156.25	Port operations
6•	156.30	156.30	Intership safety
7	160.95	156.35	International use
7A	156.35	156.35	Commercial intership and ship-to-shore
8	156.40	156.40	Commercial intership
9	156.45	156.45	General intership and ship-to-shore
10	156.50	156.50	Commercial intership and ship-to-shore
11	156.55	156.55	Commercial intership and ship-to-shore
12	156.60	156.60	Port operations
13	156.55	156.55	Navigational bridge-to-bridge
14	156.70	156.70	Port operations
15	156.75	———	Receive-only for weather information
16•	156.80	156.80	Distress, safety, and calling
17	156.85	156.85	State or local government control
18	161.50	156.90	International use
18A	156.90	156.90	Commercial intership and ship-to-shore
19	161.55	156.95	International use
19A	156.95	156.95	Commercial intership and coast-to-coast
20	161.60	157.00	Port operations
21	156.05	157.05	International use
21A	157.05	157.05	U.S. government use only
22	161.70	157.10	International use
22A	157.10	157.10	U.S. Coast Guard liaison

Table 6-2 *continued*

Channel	Shore	Ship	Use
23	161.75	157.15	International use
23A	157.15	157.15	U.S. government use only
24	161.80	157.20	Public correspondence
25	161.85	157.25	Public correspondence
26	161.90	157.30	Public correspondence
27	161.95	157.35	Public correspondence
28	162.00	157.40	Public correspondence
65	160.875	156.275	International use
65A	156.275	156.275	Port operations
66	160.925	156.325	International use
66A	156.325	156.325	Port operations
67	156.375	156.375	Commercial intership
68	156.425	156.425	Noncommercial intership and ship-to-shore
69	156.475	156.475	Noncommercial intership and ship-to-shore
70	156.525	156.525	Noncommercial intership
71	156.575	156.575	Noncommercial intership and ship-to-shore
72	156.625	156.625	Noncommercial intership
73	156.675	156.675	Port operations
74	156.725	156.725	Port operations
77	156.875	156.875	Port operations
78	161.525	156.925	International use
78A	156.925	156.925	Noncommercial intership and ship-to-shore
79	161.575	156.975	International use
79A	156.975	156.975	Commercial intership and ship-to-shore
80	161.625	157.025	International use
80A	157.025	157.025	Commercial intership and ship-to-shore
81	161.675	157.075	International use
81A	157.075	157.075	U.S. government use only
82	161.725	157.125	International use
82A	157.125	157.125	U.S. government use only
83	161.775	157.175	International use
83A	157.175	157.175	U.S. government only
84	161.825	157.225	Public correspondence
85	161.875	157.275	Public correspondence
86	161.925	157.325	Public correspondence
87	161.975	157.375	Public correspondence

• = all ships must be able to transmit and receive on this channel

A relatively new marine communications band is found at 216 to 220 MHz. This is the home of the automated maritime communications system (AMTS), which is essentially a cellular system for ship use. AMTS is used by vessels within approximately 70 miles of shore, and is currently available along the Gulf of Mexico, the Great Lakes, and major inland waterways such as the Mississippi and Ohio rivers. Within the next few years, it will expand to include the Atlantic and Pacific seaboards.

The principle behind AMTS is the same as for cellular telephones. As ships travel, their communications are switched between different shore-based cell stations. Each cell covers only a limited area, and the switching of communications between cells is done automatically. Both the AMTS cells and ship-based AMTS gear are low powered and have limited range, meaning many users can share the same frequency without interference. AMTS is currently operated by a single company—Watercom, Inc. of Jeffersonville, IN—and is available to any shipping company or vessel willing to pay for it. It is mainly used by barges and freighters on inland waterways and by off-shore oil rigs in the Gulf of Mexico. However, a few well-heeled recreational boaters also use AMTS.

Like the VHF High marine allocation, AMTS lets ships communicate with other ships, shore stations, and place telephone calls. The reliability, coverage, and quality of AMTS communications are all superior to those on the VHF marine allocations, and operations are full duplex. AMTS also offers several unique features of special interest to shipping companies. A shipping company can call a telephone number for an AMTS-equipped ship, and an on-board computer will relay such information as current location, arrival time at next port, fuel consumption, and the captain's log entries via AMTS back to company headquarters. Such data can be obtained directly from the ship's computers without any human intervention being necessary. Messages can be sent via fax or data packets to any ship if no human operators are present. Shipping companies with several vessels can send messages to their entire fleet or just to selected vessels.

AMTS stations are spaced each 12.5 kHz from 216 to 220 MHz, and narrowband FM is used. The shore cell stations are found from 216.0125 to 217.9875 MHz while ships and oil drilling platforms in the Gulf of Mexico are found from 218.0125 to 219.9875 MHz. In areas where there are television stations operating on channel 10 or 13, AMTS is restricted to 217.0125 to 217.9875 MHz and 219.0125 to 219.9875 MHz.

AMTS is still in its infancy, but it may well replace the VHF High marine allocations for most commercial vessels in a few years. The older VHF High marine band will still remain as a favorite of recreational boaters and non-fleet commercial ships.

Ham Radio

Ham radio operators don't spend all their time talking to somebody on the other side of the world by shortwave. Hams (more formally known as *amateur radio* operators) have VHF and UHF bands at 50–54 (6-meters), 144–148 (2-meters), 222–225, 420–450, 902–928, and 1240–1300 MHz. Like land mobile radio users, hams make extensive use of repeater systems to extend the operating range of mobile and hand-held units. But hams use modes, such as SSB, which are seldom otherwise used at VHF and UHF. Hams have developed a system of transmitting messages by computer, known as *packet radio*, which is used by nobody else but the military on the VHF and UHF bands. And some hams even use satellites designed and built by fellow hams to communicate at VHF and UHF.

However, the most common method of ham radio communication above 30 MHz is narrowband FM, much like the land mobile and marine radio services. The 144 to 148 MHz band (known as the 2-meter band, after the wavelength of signals at those frequencies) is the most heavily used such band, and scanning through that range will produce several active frequencies used by hams in your area. Both simplex and repeater activity is popular on 2-meters, and Table 6-3 gives some of the more popular simplex and repeater output channels. Narrowband FM is also found on the 50–54, 220–225, 420–450, and 1240–1300 MHz bands. (902 to 928 MHz is very lightly used at present.) Several of the repeater stations used by hams on these frequencies have connection facilities to the telephone system, and many hand-held units used by hams have built-in telephone dialing keypads to permit telephone calls to be placed through such repeaters.

Unlike many other users of the VHF and UHF bands, hams use *call letters* to identify their stations. These call letters are unique to each ham and are assigned by the Federal Communications Commission (and by the Department of Communications in Canada). Ham call letters consist of one or two letters, a single numeral, and one to three additional letters. The numeral indicates where the ham lived when the license was first issued. The numeral "6" in a ham's call sign indicates the ham lived in California when the license was first issued, while a "2" means the ham lived in either New York or New Jersey. Most hams are very proud of their call signs and many are better known to other hams by their call signs than by their first names. In the United States, ham call signs begin with the letters A, K, N, or W; in Canada, they usually start with VE.

The biggest use of these bands by hams is to talk to other hams in their area. Most of the communications you hear will be inconsequential chit-chat about the weather, sports, work, traffic, hobbies, current events, and other topics people normally talk about. Hams tend to be outgoing, gregarious people who are always looking for an excuse to say "hello!" to other hams by radio. However, hams get very serious and organized in their communications whenever an emergency or disaster takes place. Numerous hams are members of emergency service groups, and hold frequent drills to hone their abilities to provide communications services during crisis situations. It is surprising how often the communications facilities of police, fire, and other public service agencies become overloaded or even fail outright during emergencies, and during such times hams often are the only reliable means of radio communication still functioning. If a major disaster or emergency strikes your area—such as an earthquake, tornado, flood, or civil disturbance like the 1992 Los Angeles riots—then listening to hams on 2-meters and other VHF/UHF bands can be the best way to get the latest information about what is really going on.

Table 6-3

Commonly Used Frequencies by Hams on 2-Meters

Repeater Output Frequencies				Simplex Frequencies
145.11	145.35	146.76	147.09	146.43
145.13	145.37	146.79	147.12	146.46
145.15	145.39	146.82	147.15	146.49
145.17	145.41	146.85	147.18	146.52
145.19	145.43	146.88	147.21	146.55
145.21	145.45	146.91	147.24	146.58
145.23	145.49	146.94	147.27	
145.25	146.61	146.97	147.30	
145.27	146.64	147.00	147.33	
145.29	146.67	147.03	147.36	
145.31	146.70	147.06	147.39	
145.33	146.73			

While narrowband FM is the most popular method of voice communications, SSB and AM are also used to a lesser degree. Sporadic-E and F layer propagation take place on 6-meters much as on the VHF Low band, and some hams have contacted over 100 other countries on 6-meters. 2-meters and higher bands have plenty of tropospheric bending propagation, and many hams try to contact as many distant stations as they can when "tropo" propagation is present.

When you listen to bands like 2-meters, you may sometimes hear brief signals that have a harsh, "brapp–zzapp" sound to them. Such signals are packet radio. Packet radio uses a computer to arrange information (such as written text) into "packets," which are transmitted one after another until an entire message is sent. The entire sending and receiving process is done by a personal computer. Packet radio has some interesting advantages over other transmission modes. For one thing, packets are "addressed" to specific stations. If a packet is not addressed to a receiving station, the station will ignore that packet. This helps cut down on interference. Packet radio also has error detecting and correction capabilities. The number of bits of information in each packet is used to make up an error detection code. When a packet is received by another station, the error detection code is compared to the number of bits of information received in the packet. If there is a difference between the code and number of bits, the receiving station will send a request to the transmitting station to re-send the packet. Since these error detecting operations are handled by computer, it takes only a few milliseconds to catch an error and have a packet retransmitted. Packet messages can be automatically relayed from station to station across the country thanks to the message addressing capabilities.

Hams are permitted to transmit television signals on the 420–450 and 1240–1300 MHz bands; these are signals sent to other hams, not broadcasts intended for the general public. If you run across these ham TV signals, you'll hear a distorted "buzz" sound as you do when you scan across ordinary television channels. Hams also communicate through ham-only communications satellites. The most famous of these is the OSCAR (orbiting satellite carrying amateur radio) series. OSCAR satellites receive signals from ground-based hams on one band (like 420 to 450 MHz) and relay them back to Earth on a different band (like 144 to 148 MHz). OSCAR satellites are orbiting repeater stations, and let hams routinely communicate over thousands of miles at VHF and UHF frequencies. By the way, numerous astronauts are ham radio operators and operation from the Space Shuttle is almost a common thing on 2-meters.

Another use for ham radio is remote control of objects and models, especially in the 53 to 54 MHz range. These frequencies are less crowded than those in the 72 to 76 MHz band.

To operate a ham radio station above 30 MHz, you need a Technician class license issued by the Federal Communications Commission. To get this license, you must pass a written examination of 55 multiple-choice questions on FCC regulations, operating procedures, and basic radio techniques. (Knowledge of the Morse code is no longer a requirement for a ham license to operate above 30 MHz.) Tests are administered across the country each week by volunteer examiners, who are already-licensed hams certified by the FCC. After the examination is passed, the FCC sends a station license and call letters within a few weeks. Passing the written exam is simplified by the fact that the FCC releases a "pool" of exam questions to the public, and all exam questions are drawn from that pool. Several publishers issue study guides based on the latest questions, making it easy to prepare for the exact questions you will see on the exam.

If you get interested in the ham radio communications you hear on your scanner and decide you want to get your own ham license, HighText Publications publishes an excellent introductory book titled *All About Ham Radio*. This book discusses the basic concepts, terms, and ideas in ham radio, and supplies the necessary foundation to study for your license using one of the license study guides available from other publishers.

CHAPTER SEVEN

Military Communications and Telephone Calls

MANY LISTENERS ARE UNAWARE of the amount of U.S. military communications that can be easily heard on a scanner, or that there are other mobile telephone services besides cellular. Military communications can show up on virtually any frequency (or mode) above 30 MHz, while portions of the VHF Low, VHF High, and UHF bands were used for mobile telephone service long before there was such a thing as cellular telephony.

Military Communications

The U.S. military is a major user of the VHF and UHF bands. In fact, only TV and FM broadcasters are bigger users of the spectrum above 30 MHz. An enormous band—225 to 400 MHz—is devoted exclusively to military aviation. The U.S. military also has exclusive VHF bands at 137 to 144 and 148 to 150 MHz, and is a major user of the VHF Low band as well. In addition, the U.S. military may show up on almost any unused frequency from 30 to 76 MHz, as this is the normal tuning range of many widely used military communications systems. For example, the U.S. Air Force's Thunderbirds precision flying team has been reported on 66.90 MHz, which is in the middle of the frequency allocation for television channel 4. The U.S. Army often operates in the 56 to 57 MHz range in areas without a television station on channel 2, and the U.S. Navy often uses the 50 to 54 MHz ham radio band for operations at sea.

The military is equally diverse in the modes it uses. Narrow-band FM is common on the VHF Low band as well as the 137 to 144 and 144 to 148 MHz ranges. AM is used in the 225 to 400 MHz band. However, wideband FM is often used for voice communications, especially outside those segments. There is an increasing amount of computer data transmission and voice scrambling on military channels; these modes will have a bizarre "science fiction" sound and will be impossible to copy on your scanner. The military will sometimes use SSB, and AM may be used on frequencies where narrowband FM is normally used. In short, the U.S. military is very unpredictable in the frequencies and modes it uses!

Despite this unpredictability, some frequencies tend to be used more often than others. Table 7-1 lists some of the VHF Low band channels used by different branches of the U.S. military in recent years. Some of these are used for certain purposes, such as special warfare or armored divisions. All the frequencies in Table 7-1 are narrowband FM. Note that every service except the U.S. Air Force is represented in Table 7-1; this is because most of their communications take place in the military aeronautical band. Most of their ground-based com-munications (for such purposes as base security. etc.) take place in the 137 to 144 or 144 to 148 MHz segments.

Even if you're not located near any military base or other facilities, you can still hear plenty of military communications. The frequencies in Table 7-1 become especially active during sporadic-E or F layer propagation. You'll be able to recognize these signals by their distinctive military lingo ("Alpha Base, this is Bravo Deuce requesting switch to secure on the Bravo net") and colorful station identifiers ("Long Rifle" is used by the U.S. Marine Corps at Camp Pendleton in California). During the 1991 Gulf War, there was superb F layer propaga-tion that permitted scanner listeners in the United States to follow the progress of U.S. and other allied forces in the war with Iraq. Listeners in the United States were able to hear communications during tank battles with Iraqi forces or U.S. Navy planes conducting missions against Iraqi positions. Most listening during sporadic-E or F layer propagation won't be so dramatic, but you will hear a lot of training operations during the late spring/early summer sporadic-E "season."

Table 7-2 gives some common U.S. military aviation frequencies in the 225 to 400 MHz band. Most channels are spaced 100 kHz apart, although some (like the U.S. Air Force's 266.05 MHz) are spaced at 50 kHz intervals. Your chances of hearing Navy and Coast Guard aircraft are naturally greater if you live near the coast or a major base, while the U.S. Air Force and Air National Guard can be heard throughout the United States. The Canadian armed forces also use this band, with 245.70 and 316.50 MHz often used by the Royal Canadian Air Force.

Table 7-1

Selected VHF Low Band Military Frequencies

Frequency	Service	Frequency	Service
30.05	U.S. Army	34.40	U.S. Army (airborne)
30.15	U.S. Army (Corps of Engineers)	34.65	U.S. Marine Corps
	U.S. Navy (special warfare)	34.95	U.S. Army
30.25	U.S. Army (armored divisions)	36.15	U.S. Navy
30.30	U.S. Army	36.54	U.S. Navy
30.35	U.S. Marine Corps	36.40	U.S. Marine Corps
30.55	U.S. Army	36.60	U.S. Navy
30.85	U.S. Army (airborne)	38.25	U.S. Navy
31.05	U.S. Army (armored divisions)	38.60	U.S. Coast Guard
31.30	U.S. Army	38.70	U.S. Navy
31.60	U.S. Army (armored divisions)	39.62	U.S. Coast Guard
32.05	U.S. Navy	40.39	U.S. Coast Guard
32.10	U.S. Army (artillery)	40.79	U.S. Navy
32.20	U.S. Army (missile operations)	41.10	U.S. Navy
32.60	U.S. Army (armored divisions)	41.20	U.S. Marine Corps
32.95	U.S. Army	41.55	U.S. Marine Corps
33.35	U.S. Army	41.93	U.S. Marine Corps
33.95	U.S. Army	49.80	U.S. Navy (special
34.15	U.S. Navy		warfare)
34.30	U.S. Coast Guard		

As on the civilian aviation band, you can receive aircraft transmissions at surprising distances. Often you can hear military aircraft at greater distances than civilian aircraft since the operating altitudes of military aircraft will be greater. As a result, you may hear many "one-sided" communications where you can hear the aircraft but not the ground station it is in communication with. Fortunately, you can also hear many aircraft-to-aircraft communications, such as mid-air refueling operations, which can be heard over a wide area. However, many larger civilian airports are equipped to communicate with military aircraft and you might hear these channels come into use by your local airport if a military flight is scheduled to arrive or if an air show is scheduled.

Figure 7-1: The PRO-2006 from Radio Shack is an excellent scanner for monitoring military communications. It covers the 225 to 400 MHz military aviation band and can scan up to 400 channels in ten banks.

Like civilian aviation, there is an air traffic control system used to direct military aircraft throughout North America, known as Air Route Traffic Control (ARTCC). This network is very similar to the Unicom system used by civilian aviation. Table 7-2 lists various control tower frequencies for the different services.

The U.S. Air Force is by far the biggest user of the military aeronautical band. Table 7-2 identifies several different activities heard on various channels. The Strategic Air Command

Table 7-2

Selected Military Aviation Frequencies

Frequency	Service	Frequency	Service
225.00	U.S. Navy	317.70	U.S. Coast Guard
229.60	National Guard	317.80	U.S. Coast Guard
235.10	U.S. Air Force (air refueling)	319.40	U.S. Air Force (Military Airlift Command)
236.60	U.S. Air Force (control towers)		
240.60	U.S. Coast Guard	320.20	U.S. Navy
241.40	U.S. Navy	321.00	U.S. Air Force (Strategic Air Command)
242.20	U.S. Air Force (Tactical Air Command)		
		326.30	U.S. Air Force
250.85	U.S. Air Force	335.70	U.S. Air Force
255.40	U.S. Coast Guard	340.20	U.S. Navy (control towers)
257.80	U.S. Air Force (control towers)	340.80	U.S. Air Force (Military Airlift Command)
264.20	U.S. Navy		
266.05	U.S. Air Force (Strategic Air Command)	342.20	U.S. Coast Guard
		345.00	U.S. Navy
270.60	U.S. Navy	348.60	U.S. Air Force (control towers)
275.10	U.S. Coast Guard		
	U.S. Navy	354.20	U.S. Air Force
275.80	U.S. Air Force (control towers)	357.90	U.S. Navy
276.90	U.S. Air Force	358.90	U.S. Navy
277.80	U.S. Navy (fleet communications)	359.40	U.S. Navy
		364.20	U.S. Air Force (North American Air Defense Command)
280.50	U.S. Air Force (Tactical Air Command)		
282.70	U.S. Air Force (air refueling)	368.60	U.S. Air Force
282.80	U.S. Coast Guard (search and rescue)	370.40	U.S. Air Force
		372.80	U.S. Air Force (Military Airlift Command)
283.90	U.S. Air Force (air refueling)		
286.90	U.S. Air Force	375.70	U.S. Air Force (Strategic Air Command)
289.70	U.S. Air Force (air refueling)		
289.90	U.S. Air Force	376.20	U.S. Air Force (Tactical Air Command)
294.20	U.S. Air Force		
296.20	U.S. Air Force	378.80	U.S. Air Force
297.00	U.S. Air Force (Military Airlift Command)	380.55	U.S. Navy
		381.70	U.S. Coast Guard
300.60	U.S. Navy	382.50	U.S. Air Force (Tactical Air Command)
302.70	U.S Air Force		
305.50	U.S. Air Force	383.90	U.S. Coast Guard
306.60	U.S. Air Force (Tactical Air Command)	394.80	U.S. Air Force
		385.25	U.S. Navy
307.85	U.S. Navy	387.40	U.S. Navy
309.35	U.S. Navy	390.90	U.S. Air Force (Military Airlift Command)
310.75	U.S. Navy		
311.00	U.S. Air Force (Strategic Air Command)	395.9	U.S. Navy
		398.50	U.S. Air Force
313.60	U.S. Air Force (Tactical Air Command)		

(SAC) is the division of the Air Force responsible for undertaking offensive bombing operations against foreign targets. The Tactical Air Command (TAC) is responsible for air defense of the North American continent. On SAC and TAC channels, you'll hear many of the aircraft using tactical identifiers such as Defiance, Renegade, Repeat Lightning, and Cablecar. Tactical identifiers are changed on a daily basis. The air refueling channels, like 282.70 MHz, are used to keep aircraft aloft during lengthy missions.

The Military Airlift Command (MAC) is used to transport people and equipment between different bases. These communications can be interesting when some VIPs, such as generals or congressmen, are aboard. These frequencies were active throughout North America in late 1990 and early 1991 as troops and equipment were moved to the Middle East for the confrontation with Iraq.

The North American Aerospace Defense Command (NORAD) is a joint operation of the U.S. and Canadian Air Forces to provide early alert of any attack on North America by bombers and missiles. 364.20 MHz is NORAD's major UHF channel, and is used for all airborne intercept operations. You can hear numerous airborne electronics surveillance aircraft (known by the acronym AWACS) on this channel. North America is divided into several regions by NORAD, and separate channels are used for non-intercept operations within each region.

The U.S. Navy is another big user of aeronautical channels, and their frequencies in Table 7-2 will be especially active if you live along the Atlantic or Pacific coasts or the Gulf of Mexico. Major U.S. Navy aviation centers include Jacksonville and Norfolk on the Atlantic seaboard, Pensacola, Florida, on the Gulf of Mexico, and San Diego and Whidby, Washington on the Pacific Coast.

Non-Cellular Telephones

People had telephones in their cars long before there were cellular telephones, and cordless telephones are commonplace in many homes. It's very easy to hear these on most scanners.

Cordless telephones operate in ranges set aside for kids' walkie-talkies, remote control devices, and other miscellaneous, low-powered radio transmitting devices. Cordless telephones use duplex transmission, with base units operating in the 46 MHz range and the remote handsets found in the 49 MHz area. Both sides of a conversation can generally be heard on a base unit frequency. Table 7-3 gives the channels pairs used for modern cordless phones. Since these frequencies are limited, it's not unusual to hear two or more conversations in progress at once on a channel; a clue this is happening is badly distorted, unintelligible audio on a frequency. If you have a cordless telephone, you may have experienced interference from other cordless phones in your area that operate on the same channel pair. To help reduce interference between cordless telephones, some have a "tone access" feature similar to those used with repeater systems. The base and handset units of such cordless telephones cannot communicate with each other unless they both use the same access code. However, such cordless telephones can still be easily heard on nearby scanners. Since most cordless telephone users are unaware that their conversations can be easily heard by other people, it's vital for scanner users to respect the privacy of users and avoid listening to cordless telephone conversations.

Cordless telephones use FM. The base units use higher power than the handset units, although both use transmitter powers that are a fraction of a watt. Such low powers normally restrict the range of cordless telephones

Table 7-3

Cordless Telephone Frequencies

Base Unit	Handset
46.61	49.70
46.63	49.845
46.67	49.86
46.71	49.77
46.73	49.875
46.77	49.83
46.83	49.89
46.87	49.93
46.93	49.99
46.97	49.97

to less than 100 yards. However, it is not uncommon to hear cordless phones at far greater distances when using a scanner with a good outdoor antenna. Reception of base units at distances of up to five miles is not uncommon with good equipment, and there have even been some reports of reception at distances of several hundreds of miles during sporadic-E propagation.

Before cellular telephony, mobile telephone service was provided through a system that was much like a land mobile version of today's cordless telephones. Table 7-4 gives the base station frequencies used for older mobile telephones in the VHF Low, VHF High, and UHF bands. Table 7-4 also gives the most popular base/mobile channels, which are in the VHF High band. While most mobile telephone service today uses cellular technology and frequencies above 800 MHz, the older frequencies in Table 7-4 are still in use, particularly in rural areas.

Older mobile telephone service is duplex, with base and mobile units transmitting on separate frequencies. The base unit will transmit continuously. If no call is being handled, the base station will transmit a 2000 Hz tone, which is known as the *idle* tone. This continuous tone means most scanners will stop and "freeze" on a channel used by a base unit even if no calls are being handled. The mobile frequencies will be quiet until a mobile unit is engaged in a call. Repeater stations are not used in these mobile telephone systems, although the base unit transmitting facilities are normally located atop a mountain, tall building, or other location that gives favorable coverage.

The obvious disadvantage of older mobile telephone systems is the limited number of users that any channel can accommodate without causing interference. Another problem is that only one call can be handled on a channel at any one time. Before the advent of cellular technology, only a few hundred people could have mobile telephones in large cities like New York and Los Angeles. And those people often had to wait for the channel to clear before they could place or receive a call from their car. Users in rural areas didn't face

Table 7-4

Non-Cellular Mobile Telephone Frequencies

VHF Low Base Frequencies		UHF Base Frequencies		Commonly Used Channel Pairs		
				Channel	Base	Mobile
35.26	35.46	454.025	454.35	JL	152.51	157.77
35.30	35.50	454.05	454.375	YL	152.54	157.80
35.34	35.54	454.075	454.40	JP	152.57	157.83
35.38	35.62	454.10	454.425	YP	152.60	157.86
35.42	35.66	454.125	454.45	YJ	152.63	157.89
VHF High Base Frequencies		454.15	454.475	YK	152.66	157.92
		454.175	454.50	JS	152.69	157.95
152.03	152.57	454.20	454.525	YS	152.72	157.98
152.06	152.60	454.225	454.55	YR	152.75	158.01
152.09	152.63	454.25	454.575	JK	152.78	158.04
152.12	152.66	454.275	454.60	JR	152.81	158.07
152.15	152.69	454.30	454.625			
152.18	152.72	454.325	454.65			
152.21	152.75					
152.51	152.78					
152.54	152.81					

such problems, so today you might find the frequencies in Table 7-4 quiet in Los Angeles or New York but still in use in the Texas, Nebraska, or Maine countryside. Of course, the VHF Low band channels can be heard at great distances via sporadic-E and F layer propagation. The VHF High channels can also be heard hundreds of miles away via tropospheric bending.

Paging Signals

"Beepers" hang from the belts of all sorts of busy people. Doctors, technicians, sales people, repair and service personnel, volunteer firefighters, and even drug dealers use paging receivers so they can be quickly reached. Paging messages are transmitted to beepers on the VHF and UHF bands using the frequencies shown in Table 7-5.

Figure 7-2: Voice and data paging systems, like these units made by Courier, can be easily heard on scanners.

There are two types of paging services. One type is non-voice, and generally will only transmit the number the person with the beeper is supposed to call. If you intercept a paging signal of this type, all you will hear will be a series of tones ("bleeps" and "bloops"). The other transmits a voice message to the beeper. Both methods use a unique access tone code to activate individual paging receiver in the system. When a paging receiver receives a signal containing its access code, it makes some sort of signal that a message is arriving. This can be the "beeping" sound or some other audible alert. Newer pagers may silently alert users by vibrating.

Most pager transmissions are in FM, although AM is sometimes used, especially in Canada. Repeater signals on the VHF and UHF bands do not use repeaters, but a form of cellular technology is being used on 931 MHz. When sporadic-E or F layer propagation is taking place, paging signals on the VHF Low band can be heard from thousands of miles away. During the 1988 to 1991 peak of solar activity, a paging station on

31.35 MHz in Montevideo, Uruguay was heard throughout North America on an almost daily basis. Paging transmissions from Germany in AM were also heard regularly on 40.68 MHz during those years in the eastern half of North America; in western North America, Japanese paging services were heard on 45.36 MHz.

Table 7-5

Like other telephone services, you should respect the privacy of users of the frequencies in Table 7-5. Paging services are covered by the provisions of the Electronic Communications Privacy Act; you are prohibited from listening to any signal that you know is a paging transmitter. This is because many people who send messages to pagers by voice do so from their telephone and are unaware that what they say is transmitted over the open airwaves.

Telephone Paging Frequencies

Frequency

35.20	43.62	454.20	931.2875
35.22	43.66	454.225	931.3125
35.24	152.03	454.25	931.3375
35.26	152.06	454.275	931.3625
35.30	152.09	454.30	931.3875
35.34	152.12	454.325	931.4125
35.38	152.15	454.35	931.4375
35.42	152.18	454.375	931.4625
35.46	152.21	454.40	931.4875
35.50	152.24	454.425	931.5125
35.54	152.51	454.45	931.5375
35.56	152.54	454.475	931.5625
35.58	152.57	454.50	931.5875
35.60	152.60	454.525	931.6125
35.62	152.63	454.55	931.6375
35.66	152.66	454.575	931.6625
43.20	152.69	454.60	931.6875
43.22	152.72	454.625	931.7125
43.24	152.75	454.65	931.7375
43.26	152.78	931.0125	931.7625
43.30	152.81	931.0375	931.7875
43.34	158.10	931.0625	931.8125
43.38	158.70	931.0875	931.8375
43.42	454.025	931.1125	931.8625
43.46	454.05	931.1375	931.8875
43.50	454.075	931.1625	931.9125
43.54	454.10	931.1875	931.9375
43.56	454.125	931.2125	931.9625
43.58	454.15	931.2375	931.9875
43.60	454.175	931.2625	

CHAPTER EIGHT

A Scanning Menagerie

THE VHF AND UHF BANDS are loaded with all sorts of miscellaneous, unusual signals in addition to the more conventional ones we have examined in previous chapters. These signals include wireless microphones, baby monitors, foreign military services, and even unlicensed, illegal land mobile operations. This chapter is a look at the "wild side" of scanner monitoring.

Wireless Microphones

The Federal Communications Commission allows a variety of low-power transmitting devices to operate on frequencies normally assigned to other services. The FCC rules covering such devices restrict them to low power, usually only a fraction of a watt, so they are unlikely to cause any interference to other users of the frequency. The maximum range of such devices is usually less than 100 yards. Wireless microphones are common examples of these low power units.

One very common range for wireless microphones is the 88 to 108 MHz FM broadcast band. Such "mikes" can be tuned to a clear frequency in the FM band, and are often small enough to clip onto a jacket lapel or even a tie. These units have been sold for years by Radio Shack and other electronics stores, and were popular because of their low price. As the FM broadcast band got more crowded, however, it became increasingly difficult to find a clear spot for these mikes. Manufacturers of FM wireless microphones started making units for other frequencies, and Table 8-1 gives some of the more common frequency ranges used today. The most popular frequency range for low cost wireless microphones is 47.02 to 47.50 MHz. Another popular frequency range is 72 to 76 MHz. Wireless microphones that use that range do so in 25 kHz increments beginning at 76.025 MHz. More expensive professional-quality wireless microphones often use frequencies in the VHF High band and above. Some of the more popular frequencies for these units are 169.55, 170.245, 171.905, 174.60, 174.80, 175.00, 177.60, 180.60, 181.60, 183.60, 184.00, 186.60, 190.60, 192.60, 194.60, 195.60, 196.60, 199.60, 202.40, and 203.40 MHz. However, they often show up anywhere between television channels 7 to 13 if no local stations use one of those channels.

The limited range of wireless microphones does not prevent a lot of interesting listening, particularly if you use a portable hand-held scanner at a public event. Many concerts, speeches, and other events make use of multiple wireless microphones. If you have ever seen a concert where the singers and musicians move about as they perform and no cables or wires are visible, then you're seeing wireless microphones in action. The backstage receivers for these microphones are not connected to the auditorium sound system continuously, although the microphones are often left on all the time. As a result, a hand-held scanner at a concert can let you hear the musicians tuning their instruments or warming up prior to the show. Vocalists can be heard getting their voice ready or—more often than you might expect—making some choice comments about the audience, band members, the auditorium, or how they're feeling that night. Sometimes a wireless mike will be left on a table or in a dressing room before the performer takes the stage, and in such circumstances the mike will be like an eavesdropping device, letting you hear everything that is said or done in the area.

Table 8-1

Wireless Microphone Frequencies in MHz

47.02 to 47.50
72.00 to 76.00
88.00 to 108.00
157.05 to 157.11
169.05 to 171.95
174.50 to 177.90
180.20 to 186.90
190.20 to 199.90
202.10 to 203.90

A scanner can let you enjoy a concert for free if the show is being held in an outdoor setting, such as a concert bowl or stadium. It's possible to sit in the parking lot or other area adjacent to the concert site and listen to the entire show over your scanner.

Eavesdropping on wireless microphones made headlines a few years ago. James Randi, a stage magician under the professional name of The Amazing Randi and an investigator of claims of paranormal experiences, became intrigued at the alleged faith healing activities of Peter Popoff, a California television evangelist. One of the most famous parts of each program would involve Popoff's seemingly miraculous ability to diagnose the illnesses of audience members, even down to the names of their doctors and the course of treatment previously undertaken. During his investigation, Randi noted that members of Popoff's staff would engage audience members in friendly, casual conversation before each show. During these chats, staff members would learn details of the audience members' lives and illnesses. Popoff wore an earphone during his sessions with audience members, and Randi suspected that the information compiled by his staffers was being fed back to Popoff by radio. Using a portable scanner, it was discovered that Popoff was indeed receiving data from his staffers on 39.17 MHz. Aides (including Popoff's wife) backstage directed Popoff toward certain audience members and recited various information for Popoff to repeat. Randi describes the entire incident is his book *The Faith Healers*.

Similar listening fun can be had when speakers are using wireless microphones at events such as political rallies. Listeners in New York City

Figure 8-1: A portable scanner, like the Bearcat BC70XLT from Uniden, can let you listen to wireless microphones and other "behind the scenes" happenings at concerts, conventions,

reported some interesting "behind the scenes" remarks by speakers at the 1992 Democratic National Convention held at Madison Square Garden. Several of the speakers, including then-Governor Bill Clinton, used wireless microphones to deliver speeches. These microphones were "on" before and after speeches, and scanner-equipped listeners in the Garden were able to hear one national political figure repeat "don't screw up! don't screw up!" to himself before he took the podium for a nationally televised speech during the convention.

Walkie-Talkies

Small, inexpensive walkie-talkies popular as kids' toys and more expensive units with remote headsets both operate near the upper end of the VHF Low band. Commonly used frequencies include 49.83, 49.845, 49.86, 49.875, and 49.89 MHz. All of these units use FM and very low power, only a fraction of a watt. The range of these units is less than a half-mile under most circumstances, although a scanner with an outdoor antenna can receive such devices at distances of two or three miles.

Normally, kids playing with their walkie-talkies will be more of an annoyance than anything else. This is especially so if you do a frequency search and your scanner stops on a frequency where a couple of kids are yelling back and forth to each

Figure 8-2: Walkie-talkies like this operate on frequencies around 49 MHz.

other. However, the more expensive units with headsets are often used by businesses and even police and fire departments for short-range communications.

Room Monitors and Wireless Intercoms

The same FCC rules that allow low power wireless microphones and walkie-talkies have also let manufacturers offer "wireless intercoms" and "room monitors" (the latter are also sold as "baby monitors"). The wireless intercoms are little more than walkie-talkies with squelch circuitry in their receiving sections, while room monitors are, for all practical purposes, eavesdropping devices that people voluntarily install in their homes.

FCC regulations permit these devices to operate anywhere in the 49.82 to 49.90 MHz range, although most units operate on 49.83, 49.845, 49.86, 49.875, or 49.89 MHz. Some units come equipped with a "slider" control to slightly adjust the transmitting and receiving frequencies so interference can be avoided. As with other low power devices, these can be heard far beyond their stated normal range if a scanner and outdoor antenna are used.

Since they're used intermittently, wireless intercoms are hard to deliberately listen to. However, if your scanner searches the 49.82 to 49.90 MHz range and locates a sporadic signal there, it may well be a wireless intercom in your neighborhood. On the other hand, room monitors operate continuously. If you do a frequency search of their operating range, your scanner will probably stop and "freeze" on any frequency where one is operating. You may hear no audio whatsoever on the frequency, but you're more likely to hear a weird assortment of background sounds, whimpering or crying noises, and echoes from other rooms. The effect is much like pressing your ear against a wall and listening to what's going on in the next room.

Listening to room monitors is not covered by the Electronic Communications Privacy Act, nor is it forbidden by any existing state statutes. While legal, it does not seem very ethical. People who buy and install room monitors usually have no clue that they are, in effect, bugging their own homes; many think the system uses house wiring or some frequencies that can't be received on a scanner for operation. Users of these devices are not aware they are letting people as far as two or three miles away listen to what's going on inside their homes. The courteous and decent thing to do is to tune away from any room monitors you may come across; even if a window is open, that doesn't mean you have to look in. If you own and use a room monitor yourself, be aware that other people can easily hear its signals. This can have some practical benefits; for example, you can use a hand-held scanner to listen to the room monitor while you're working in the yard.

Illegal Communications

If you live in a major urban area, particularly metro New York or southern California, you will soon become aware that there are many users of the VHF and UHF bands who have never bothered to get a license from the FCC and seem to operate as they please. And even if you don't live in a big city, you will still run across some "funny stuff," especially when sporadic-E propagation is taking place.

One of the curious aspects of American law is that it is illegal to use most radio transmitting equipment, except for CB radios and low power devices such as the ones discussed earlier in this chapter, without the proper license from the FCC. However, there is no prohibition against owning such equipment without a license or selling equipment to people without a license. While the number of legal (and illegal) radio transmitters has skyrocketed in recent years, the FCC's enforcement budget has actually been reduced over the same period and fewer personnel are available to police the radio bands than there were a few years ago. The result has been—not surprisingly—that people are beginning to buy two-way radio equipment and start using it on frequencies and in ways other than those permitted by the FCC.

Back in Chapter Four we looked at some of the legal users of the 29.7 to 30 MHz range. If you've done much listening in that segment, you've no doubt noticed that the illegal users sometimes seem to outnumber the legal users. Among the more common "bootleg" communications you will hear (especially during sporadic-E propagation) will be fishing boats. These transmissions will usually be in AM or SSB, and conversations will range from the day's catch and ocean conditions to graphic descriptions of various sexual activities. You can also hear hobbyists conducting two-way communications similar to ham radio operators (even down to what seem like authentic call signs).

The 29.7 to 30 MHz segment and the VHF Low band are also populated by several unlicensed taxicabs, limo, and delivery services. Many of these businesses are operated by recently arrived (and often illegal) immigrants, so you can hear such exotic foreign languages as Vietnamese, Arabic, Farsi, Russian, and Cambodian over the radio channels they use. In the New York and Los Angeles metropolitan areas, there are many "gypsy" taxis using these frequencies to provide transportation in ethnic neighborhoods or for people who speak a language other than English. (The term "gypsy" comes from the fact that the taxis have no city license and, for that matter, the drivers seldom have a valid driver's license.) You can tell a gypsy cab by the references to local airports and landmarks (like "LaGuardia" or "Century City") amid the chatter in foreign languages. Most of these operations take place below 36 MHz and use FM, although some AM might be heard (some operators use modified CB units). Even if you don't live near New York or Los Angeles, you can often hear these communications when the annual sporadic-E season arrives in late

May. Why doesn't the FCC go after these (and other) illegal communications networks? Besides a shortage of people and funds, there is also the matter of where these stations are and who operates them. Many are located in rougher, more danger-ous areas of New York and Los Angeles, and the people who operate them can be rough, dangerous individuals. They are not the sort of areas or people you want to encounter if you're an unarmed FCC inspector!

Some "bootleggers" on the VHF Low band are relatively benign, such as the interstate truckers that use 31.10 and 37.00 MHz to escape the crowds on the CB channels. There are several legitimate, otherwise law-abiding individuals and busi-nesses that use unlicensed two-way radio equipment simply because they are unaware that a license is required. (This is often because they assume all two-way radio equipment is like CB radio, which doesn't require a license.) Others are not so innocent. Drug smugglers and dealers make extensive use of the VHF Low and High bands, with the itinerant channels (see Chapter Five) often used by local dealers. Smugglers are mainly heard in coastal and border regions, while dealers can be heard virtually anywhere. Drug-related communications in the United States and Canada tend to be veiled and a bit vague; tip-offs can be delivery instructions for shipments that seem a bit odd—how many people legitimately make a delivery of "brownies" to a nightclub parking lot at midnight? Some surprisingly candid drug-related communications can be heard through sporadic-E propagation. These are mainly from the Caribbean area, and usually involve rendezvous and transfer points for drug shipments. Via sporadic-E reception, you can hear explicit references to different types of drugs and various locations along the shorelines of different Caribbean islands.

Other illegal users of the VHF and UHF bands include prostitution rings, "coyotes" who bring illegal immigrants into the United States and move them into major urban areas (these include urban areas like Kansas City and Atlanta, which are far removed from the Mexican border), and urban gangs. The latter are especially fond of using hand-held units operating on itinerant channels. Marine VHF transceivers are also becoming popular for illegal use in areas away from major bodies of water. The marine channels are empty and largely unmonitored by official agencies in many inland areas, and illegal operations on obscure channels far from the water stand a good chance of going unnoticed. Even if you're located in the middle of Nebraska or North Dakota, it might pay to scan through the VHF marine channels from time to time.

Even commercial land mobile radio users have problems with illegal operators. In larger metropolitan areas, there is a booming black market in two-way radio equipment having the necessary tone access encoder to use shared repeater systems. The May 27, 1992 issue of the Los Angeles *Times* carried a story about such interference plaguing a shared repeater network owned and operated by Van Williams, who played the "Green Hornet" in the 1960s television series. For several months, Williams and the users of his repeaters had been bothered by a group of operators who had managed to obtain the access codes for his repeaters and had installed the necessary tone encoding devices in their equipment. Several requests to the FCC for help in locating the illegal operators went unheeded, so Mr. Williams enlisted the Los Angeles police department in tracking down the bootleggers. Mr. Williams employed an electronics technician for help in locating the source of the signals on the input frequency of his repeaters, and when the source was identified the police arrested the operators for theft of services (the use of the repeater without paying the monthly fee). It turned out that the bootleggers had been using equipment they had bought at a local electronics swap meet.

Given the steadily growing tide of lawlessness in the United States and the FCC's proven inability to police the radio bands, it seems certain that more illegal and bootleg users of the VHF and UHF band will appear in the future.

Space Communications

The Space Shuttle uses 259.70, 279.00, and 296.80 MHz in the military aeronautical band for communications during the launch and landing portions of each flight. (During the flight itself, the Shuttle uses communications satellites for reliability.) These channels are used from lift-off until about six minutes into a flight and again from the moment the Shuttle re-enters the lower stratosphere until it lands at either Edwards Air Force Base in California or at Cape Canaveral in Florida. Since communications are restored on re-entry while the Shuttle is at a high altitude, the Shuttle can usually be heard throughout southern California and northern and central Florida when it lands. Like other military aeronautical band communications, Space Shuttle communications are in AM.

The Russian (formerly Soviet) manned space program has used 143.625 MHz for its communications since the very first flight into space by Yuri Gagarin in 1961. This frequency is currently used by the Mir space station and the Soyuz vehicles that ferry cosmonauts from Earth and back. This channel is wide-band FM, and sometimes transmits continuously while cosmonauts are aboard Mir, much like a room monitor. Two other active channels for Russian manned spacecraft are 121.75 and 142.417 MHz, which are often used for communications between Mir and Soyuz vehicles during the arrival or departure of a Soyuz. Since Soyuz and Mir operate in orbit at an elevation of approximately 200 miles, they can be heard over a radius of thousands of miles. However, Mir and Soyuz are in motion as they orbit, and this produces two phenomena that affect listening. One is that the Soyuz or Mir moves in and out of reception range of the location where you are listening. Most of the time, 143.625 MHz will be empty of any signals from Mir or Soyuz. But as they move within range of your position, the channel may abruptly "come alive" with communications from the Russian vehicles. When they move out of range, signals will "break off" just as suddenly. The period during each orbit when Mir and/or Soyuz will be audible is known as a *pass*. A second effect is known as the Doppler shift, and means the frequency of the Russian signals will seem to drop during each reception. The effect is similar to the way the pitch of an ambulance siren will seem to drop as it moves away from you. On your scanner, this means a signal from Mir or Soyuz will appear to be slowly drifting down in frequency as you listen, and there may be some distortion as a result.

International Reception

Earlier in this book, we have mentioned how stations on the VHF Low band can be received at incredible distances during periods of sporadic-E and F layer propagation. Even if you don't go looking for such unusual radio receptions, the odds are good that you will run across some sporadic-E propagation each summer if you do any listening at all on the VHF Low bands. And what you can hear during years of high sunspot activity is really incredible.

The peak of the last sunspot cycle took place in 1988 to 1991. During the winter months of those years, scanner listeners in North America were treated to the sort of reception that normally requires a shortwave radio. Listeners in eastern North America were able to hear stations in Europe, Africa, and the Middle East while scanner fans in western North America were treated to stations from Asia, the Pacific, and Australia. Perhaps luckiest of all were those in central North America, who were able to hear all of those places at one time or another. Users of the VHF Low band also experienced the same unusual propagation, but they were much less happy about it. Businesses and police departments in the United States sometimes found their repeaters relaying stations speaking French,

Spanish, German, and even Turkish. Some of these stations had operators that spoke some English, and they often delighted in hassling U.S. stations

During Operation Desert Storm in early 1991, numerous scanner listeners in eastern and central North America were able to intercept communications directly from Iraq, Kuwait, and Saudi Arabia on the VHF Low band. Some lucky listeners were able to "ride along" with the first Allied tank convoys to enter Kuwait and to also listen to bombing missions against fleeing Iraqi forces. Some listeners intercepted what was later identified as the countdown to the launch of a Scud missile—and the countdown and launch procedure was conducted in Russian, presumably by Iraqi personnel trained by the former Soviet Union.

Most receptions due to F layer propagation are much less spectacular, however. During the last sunspot cycle peak, listeners in eastern North America frequently heard German autobahn patrols, Dutch paging stations, police departments in Finland, and Romanian government stations. Various police departments in South Africa were heard with surprising regularity on such channels as 32.10 and 38.00 MHz; some listeners were able to hear communications directly from the scene of demonstrations against the South African government's apartheid policies; many of these turned violent as scanner users in the United States listened. Western North America listeners heard plenty of South Korean military operations around 34 MHz, public utility departments in Australia, Japanese stations of every description, and even such esoterica as cable television companies on Guam dispatching installation personnel.

When F layer propagation is possible, distant stations to the east start fading in shortly after sunrise at your location. As the day progresses, stations to the east start fading out and most European and African stations are gone shortly after your local noon. F layer reception during the early afternoon is concentrated on southern South America, with stations from Chile,

Uruguay, Argentina, and southern Brazil becoming dominant. By late afternoon, stations from the west begin fading in. Hawaiian stations are the first to become audible by mid-afternoon, followed by stations in Asia and Australia. Reception from the west is best in the last two hours before sunset each day. At sunset, F layer reception rapidly begins to fade and the VHF Low band is usually back to normal within a half-hour after sunset. To get the most out of F layer propagation, a good outdoor antenna, preferably one that is directional and rotatable, is a must. Unfortunately, the golden days of F layer propagation won't return until toward the end of this century.

Sporadic-E reception happens every year regardless of the sunspot cycle, and can bring in a wealth of foreign stations up to about 1500 miles from your location. A knowledge of Spanish will be a big help in making sense of what you're hearing, as stations in Central America and northern South America are heard in North America each year from late May through July. Not everything will be in Spanish; the Bahamian Coast Guard, bookie operations in Jamaica, and resorts in the Virgin Islands all use English and are well heard. French is sometimes heard from such places as Martinique and French Guiana. Often one of the first clues that sporadic-E propagation is taking place is the sudden appearance of voices speaking Spanish on the output frequencies of local VHF Low band repeaters!

Surveillance and Undercover Operations

Don't think that you're ready to hear all your local police communications because you have all the frequencies listed for them in a published directory entered into your scanner. If you live in a medium-sized or large city, odds are that the police use several unpublished frequencies for surveillance, undercover, and stakeout operations.

There's nothing sinister about this. The Federal Communications Commission allows police departments to use units of two watts or less of power on any public safety (police and fire departments), local government, highway maintenance, or conservation frequency from 42 to 952 MHz. In addition to surveillance and undercover operations, police use low power units on these frequencies for hidden microphones (including those worn on the body) and tracking transmitters concealed on a vehicle. In addition, some police departments use 49 MHz walkie-talkies, especially the "hands free" type that combine an earpiece and microphone into one, for surveillance work. There have even been some reports of police departments using the VHF marine channels for surveillance and stakeout activity, although it is illegal for them to do so.

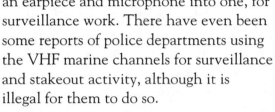

Figure 8-3: For searching out surveillance devices, a wide coverage scanner with multiple memories—like this deluxe model from Shinwa—is necessary.

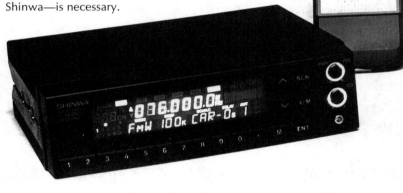

It might be difficult to determine if you're listening to surveillance or undercover operations by a law enforcement agency or some sort of illegal or bootleg activities. Transmissions will tend to be cryptic and not make a great deal of sense. Identifiers will often be of the tactical variety like those used by the military. The best clue that you're hearing something legitimate and not illegal will be the greater discipline used by law enforcement agencies; chatter will be kept to a minimum and there will be no whistling, sound effects, or other sloppiness.

"Beeper" transmitters, like their name says, transmit a single "beep" sound at periodic intervals to allow tracking of a vehicle or other movable object (like a piece of luggage). Beeper transmitters are limited to one watt and can operate on the frequency ranges given in Table 8-2. In addition, 40.22 MHz has been a popular beeper transmitter channel used by the FBI and other federal agencies. These transmitters must be limited to no more than 10 days continuous operation (usually this is accomplished by a limited battery life), so any signal heard for weeks on end is not a beeper transmitter.

Body microphones are known as "wires" in law enforcement terminology. The FM broadcast band of 88 to 108 MHz is also popular with these body mikes, and various federal law enforcement agencies like to operate "wires" in the 162 to 174 and 406 to 420 MHz ranges. "Wires" tend to use wideband FM instead of narrow-band, so place your scanner in that mode if it is available.

Needless to say, if you do intercept any genuine surveillance, undercover, or stakeout communications, the operation and officers involved will be in your neighborhood, perhaps even as close as next door. In such circumstances, it is only prudent to avoid doing or saying anything that might indicate to anyone that you are aware something unusual is going on.

Table 8-2

Beeper Transmitter Frequencies

30.85 to 32.00
33.00 to 33.07
33.41 to 34.00
37.00 to 37.43
37.89 to 38.00
39.00 to 40.00
42.00 to 42.91
44.61 to 46.60
47.00 to 47.41
150.995 to 151.490
153.740 to 154.445
154.635 to 155.195
155.415 to 156.250
158.715 to 159.465
453.0125 to 453.9875
458.0125 to 458.9875
462.9375 to 462.9875
465.0125 to 460.5125
465.5625 to 465.6375
467.9375 to 467.9875

Telemetry

You may run across several baffling "ping," "chirp," "click," and similar sounds on your scanner, particularly on the VHF Low band. These are the sounds produced by a variety of telemetry devices, which transmit a continuous stream of data. For example, telemetry buoys are used in some bays and inlets to keep track of such things as currents, tidal levels, and ocean temperature. Other telemetry transmitters are used to transmit weather data (wind speed, temperature, etc.) from remote locations such as mountaintops. Another use is in wildlife tracking; these are the "radio collars" placed around some animals taken from and later returned to the wild. The transmitters for such purposes are typically less than a watt and can show up on almost any non-public safety frequency. If you run across a strange signal that lasts for several days, and doesn't appear to be an image or other spurious signal within your scanner, then you may be hearing a telemetry signal. In Appendix C, common frequencies for these devices are indicated by "industrial, scientific, and medical devices." These are also common on the 72 to 76 MHz and other frequencies reserved for remote control purposes.

References for Scanner Listeners

T HE FOLLOWING ARE frequency directories and station guides of value to scanner listeners. Most are available from scanner dealers and radio supply houses.

Books

Police Call This is a directory of all public safety, federal government, local government, aeronautical, emergency, railroad, and highway maintenance channels from 25 to 941 MHz. It is published in several volumes organized by geographic area (for example, volume 1 covers the New England states) and is revised annually. This is perhaps the most valuable single reference for scanner listeners.

Top Secret Registry of U.S. Government Radio Frequencies Revised periodically, this is a guide to thousands of frequencies used by U.S. government agencies across the country, including the military. Also includes some Canadian government frequencies. None of the frequencies are really "top secret," although this is about the only place you can find most of them.

Tune In On Telephone Calls The title says it all. Even if you aren't interested in listening to cellular and cordless telephone conversations, this will be of interest to anyone who ever talks over such telephones.

Monitor America This is an 800+ page guide to public service frequencies across America. It also includes many business, railroad, and aeronautical channels.

Directory of North American Military Aviation Communications If you get hooked on monitoring the 225 to 400 MHz range, you'll want this book. It is published in four volumes by geographical region, and includes Canada, Mexico, and Central America in addition to the United States.

Air Scan Includes civilian, federal, and military frequencies used for aeronautical communications in the United States and Canada; essential if you're serious about aeronautical monitoring.

Magazines

Monitoring Times (140 Dog Branch Road, Brasstown, NC, 28902) Extensive coverage of military, federal government, and aeronautical scanning. Monthly scanning column along with equipment reviews and technical articles relating to scanner reception. Available at scanner and ham radio equipment dealers and some electronics supply stores.

National Scanning Report (P. O. Box 291918, Kettering, OH, 45429) Bimonthly publication covering scanner activity exclusively. Includes equipment reviews and latest frequencies in use around the country. Currently by subscription only.

Popular Communications (76 North Broadway, Hicksville, NY, 11801) Has a monthly column on scanning along with equipment reviews and feature articles. Excellent coverage of such topics as F layer and sporadic-E propagation, electronic eavesdropping techniques, and legal issues relating to scanning. Available at many newsstands.

U.S. Scanner News (706 W. 43rd St., Vancouver, WA, 98660)
Monthly publication on scanning topics, including equipment
reviews, performance modifications, and latest frequency allocations. Currently by subscription only.

Clubs

The following clubs are all-volunteer organizations operated
on a non-profit basis. As a result, it's possible that some of the
addresses below will no longer be valid or that some of the clubs
will be out of business by the time you read this. However,
there is no better way to keep up with the latest frequencies
being used in your area than membership in a local scanning
club. Always enclose a self-addressed stamped envelope when
asking for membership information; most will send a sample of
their bulletin for $2.00.

All Ohio Scanner Club, 50 Villa Road, Springfield, OH,
45503–1036. Covers Ohio and surrounding states.

Bay Area Scanner Enthusiasts, 4718 Meridian Ave., #265,
San Jose, CA, 95118. Covers San Francisco Bay area.

Houston Area Scanners and Monitoring Club, 909 Michael,
Alvin, TX, 77511. Houston area and its suburbs.

Metro Radio System, P. O. Box 26, Newton Highlands, MA,
02161. New England area.

MONIX, 7917 Third St., West Chester, OH, 45069–2212.
Cincinnati and Dayton areas.

Northeast Scanner Club, P. O. Box 62, Gibbstown, NJ, 08027.
They define "Northeast" as everything from Maine through
Virginia.

Radio Communications Monitoring Association, P. O. Box 542, Silverado, CA, 92676. National in scope, covering all activity above 30 MHz. This is the one to join if there is no club covering your area.

Radio Monitors of Maryland, P. O. Box 394, Hampstead, MD, 21074. Maryland only.

Regional Communications Network, P. O. Box 83, Carlstadt, NJ, 07072–0083. Metropolitan New York area.

Rocky Mountain Monitoring Enthusiasts, 11391 Main Range Trail, Littleton, CO, 80127. Denver area and other regions of the Mountain West.

Toledo Area Radio Enthusiasts, 6629 Sue Lane, Maumee, OH, 43537. Covers northern Ohio and southern Michigan.

Triangle Area Scanner Group, P. O. Box 28587, Raleigh, NC, 27611. Raleigh, Durham, and Chapel Hill areas of central North Carolina.

"10-Code" Meanings

MANY LAW ENFORCEMENT and public safety agencies use the "10-code" to convey ideas more quickly and consistently than can be done by using plain language. The following list is the one developed by the Association of Police Communications Officers, and is the most widely used version of the 10-code. While local agencies may vary from the standard code somewhat (usually by changing the meanings of some of the less often used codes to reflect individual agency needs), most of the 10-codes you hear on your scanner will closely follow those below.

10-1	Signal weak	10-18	Urgent
10-2	Signal good	10-19	Contact. . .
10-3	Stop transmitting	10-20	Location
10-4	Affirmative; okay	10-21	Call — by phone
10-5	Relay to. . .	10-22	Disregard
10-6	Busy	10-23	Arrived at scene
10-7	Unit going out of service	10-24	Assignment completed
10-8	Unit returning to service	10-25	Report to. . . .
10-9	Repeat last transmission	10-26	Estimated arrival time
10-10	No	10-27	License/permit information
10-11	— on duty	10-28	Ownership information
10-12	Stand by	10-29	Records check
10-13	Existing conditions	10-30	Danger/caution
10-14	Message	10-31	Pick up. . . .
10-15	Message delivered	10-32	— units
10-16	Reply to message	10-33	Emergency
10-17	En route	10-34	Time is. . . .

"10-Code" Meanings

Scanner Frequencies Used Nationwide

Here are some frequencies commonly used across the United States; some are also used in Canada. All transmissions will be in narrowband FM, or in AM from 108 to 136 MHz and 225 to 400 MHz, unless otherwise indicated. This list is a only a small sample of the frequencies used above 25 MHz, and is restricted to the most active and widely heard channels. Not all of these frequencies will be active in your area, and many users will be on additional frequencies other than those listed below. However, this list is a good starting point for your scanning. The VHF Low band listings are useful for identifying distant stations heard during sporadic-E and F layer propagation.

Frequency	Usage	Frequency	Usage
25.02	Petroleum radio service	25.87	Broadcast auxiliary
25.04	Petroleum radio service	25.91	Broadcast auxiliary
25.06	Petroleum radio service	25.95	Broadcast auxiliary
25.08	Petroleum radio service	25.99	Broadcast auxiliary
25.10	Petroleum radio service	26.03	Broadcast auxiliary
25.12	Petroleum radio service	26.07	Broadcast auxiliary
25.14	Petroleum radio service	26.09	Broadcast auxiliary
25.16	Petroleum radio service	26.11	Broadcast auxiliary
25.18	Petroleum radio service	26.13	Broadcast auxiliary
25.20	Petroleum radio service	26.15	Broadcast auxiliary
25.22	Petroleum radio service	26.17	Broadcast auxiliary
25.24	Petroleum radio service	26.19	Broadcast auxiliary
25.26	Petroleum radio service	26.21	Broadcast auxiliary
25.28	Petroleum radio service	26.23	Broadcast auxiliary
25.30	Petroleum radio service	26.25	Broadcast auxiliary
25.32	Petroleum radio service	26.27	Broadcast auxiliary

Frequency	Usage	Frequency	Usage
26.29	Broadcast auxiliary	27.315	CB channel 31 [AM or SSB]
26.31	Broadcast auxiliary	27.325	CB channel 32 [AM or SSB]
26.33	Broadcast auxiliary	27.335	CB channel 33 [AM or SSB]
26.35	Broadcast auxiliary	27.345	CB channel 34 [AM or SSB]
26.37	Broadcast auxiliary	27.355	CB channel 35 [AM or SSB]
26.39	Broadcast auxiliary	27.365	CB channel 36 [AM or SSB]
26.41	Broadcast auxiliary	27.375	CB channel 37 [AM or SSB]
26.43	Broadcast auxiliary	27.385	CB channel 38 [AM or SSB]
26.45	Broadcast auxiliary	27.395	CB channel 39 [AM or SSB]
26.47	Broadcast auxiliary	27.405	CB channel 40 [AM or SSB]
26.965	CB channel 1 [AM or SSB]	27.43	Business radio service
26.975	CB channel 2 [AM or SSB]	27.45	Business radio service
26.985	CB channel 3 [AM or SSB]	27.47	Business radio service
27.005	CB channel 4 [AM or SSB]	27.49	Business radio service
27.015	CB channel 5 [AM or SSB]	27.51	Business radio service
27.025	CB channel 6 [AM or SSB]	27.53	Business radio service
27.035	CB channel 7 [AM or SSB]	27.545	U.S. Navy/U.S. Army
27.055	CB channel 8 [AM or SSB]	27.55	U.S. Navy/U.S. Army
27.065	CB channel 9 [AM or SSB]	27.555	Drug Enforcement Administration
27.075	CB channel 10 [AM or SSB]		
27.085	CB channel 11 [AM or SSB]	27.565	U.S. Army/Department of Energy
27.105	CB channel 12 [AM or SSB]		
27.115	CB channel 13 [AM or SSB]	27.575	Numerous U.S. government agencies
27.12	Industrial, scientific, and medical devices	27.585	Numerous U.S. government agencies
27.125	CB channel 14 [AM or SSB]	27.595	Department of Energy
27.135	CB channel 15 [AM or SSB]	27.60	U.S. Army
27.155	CB channel 16 [AM or SSB]	27.605	Department of Energy
27.165	CB channel 17 [AM or SSB]	27.61	Department of Commerce
27.175	CB channel 18 [AM or SSB]	27.615	Department of Energy
27.185	CB channel 19 [AM or SSB]	27.625	Department of Energy/Federal Aviation Administration
27.205	CB channel 20 [AM or SSB]		
27.215	CB channel 21 [AM or SSB]		
27.225	CB channel 22 [AM or SSB]	27.63	National Aeronautics and Space Administration
27.235	CB channel 24 [AM or SSB]		
27.245	CB channel 25 [AM or SSB]	27.635	Department of Energy
27.255	CB channel 23 [AM or SSB]	27.645	Department of Energy
27.265	CB channel 26 [AM or SSB]	27.65	U.S. Navy/U.S. Army
27.275	CB channel 27 [AM or SSB]	27.655	U.S. Air Force/Department of Energy
27.285	CB channel 28 [AM or SSB]		
27.295	CB channel 29 [AM or SSB]	27.665	Canadian civil defense [AM mode]
27.305	CB channel 30 [AM or SSB]		

Frequency	Usage	Frequency	Usage
27.675	Department of Energy	30.09	U.S. Army
27.70	U.S. Navy/U.S. Army/ Department of Energy	30.15	U.S. Navy
		30.22	Paging signals [Canada]
27.725	National Aeronautics and Space Administration	30.30	U.S. Army
		30.35	U.S. Marine Corps
27.74	Miscellaneous federal government agencies and contractors	30.42	Paging signals [Canada]
		30.45	U.S. Army
		30.64	Special industrial radio service
27.75	U.S. Army		
27.775	U.S. Air Force	30.84	Business radio service [low power operations]
27.78	U.S. Army		
27.79	U.S. Army	31.28	Special industrial radio service
27.80	U.S. Army/U.S. Navy/ Canadian civil defense [AM mode]	31.32	Special industrial radio service
27.81	U.S. Army	31.36	Special industrial radio service
27.82	U.S. Army		
27.83	U.S. Navy	31.40	Special industrial radio service
27.85	U.S. Army/U.S. Navy		
27.87	U.S. Air Force/Strategic Air Command	31.42	Paging signals [Canada]
		31.44	Special industrial radio service
27.90	Numerous U.S. government agencies	31.48	Special industrial radio service
27.95	U.S. Army/U.S. Navy		
27.98	U.S. Air Force/U.S. Coast Guard	31.52	Special industrial radio service
29.60	Simplex ham radio communications	31.56	Special industrial radio service
29.62	Ham radio repeater output channel	31.60	Special industrial radio service
29.64	Ham radio repeater output channel	31.64	Special industrial radio service
29.66	Ham radio repeater output channel	31.68	Special industrial radio service
29.68	Ham radio repeater output channel	31.72	Special industrial radio service
29.71	Forestry radio services	31.76	Special industrial radio service
29.73	Forestry radio services		
29.75	Forestry radio services	31.80	Special industrial radio service
29.77	Forestry radio services		
29.79	Forestry radio services	31.84	Special industrial radio service
30.02	Paging signals [Canada]		
30.05	Department of Commerce	31.88	Special industrial radio service

Frequency Usage

31.92	Paging signals [Canada]
31.92	Special industrial radio service
31.96	Special industrial radio service
32.10	U.S. Army
32.42	Paging signals [Canada]
32.70	National Guard
32.85	U.S. Marine Corps
33.12	Special industrial radio service
33.14	Business radio service [low power operations]
33.18	Petroleum radio service
33.38	Petroleum radio service
33.40	Business radio service [low power operations]
33.42	Paging signals [Canada]
33.44	Fire departments
33.70	Fire departments
33.90	Fire departments
34.00	U.S. Army
34.03	Department of Energy
34.25	Fish and Wildlife Service
34.30	U.S. Coast Guard
34.65	U.S. Marine Corps
34.80	U.S. Marine Corps
34.90	National Guard/U.S. Army
34.98	Department of Commerce
35.04	Itinerant users
35.20	Paging signals
35.22	Paging signals
35.24	Paging signals
35.26	Mobile telephones/paging signals
35.28	Special industrial radio service
35.30	Mobile telephones/paging signals
35.32	Special industrial radio service
35.34	Mobile telephones/paging signals

Frequency Usage

35.36	Special industrial radio service
35.38	Mobile telephones/paging signals
35.40	Special industrial radio service
35.42	Mobile telephones/paging signals
35.44	Special industrial radio service
35.46	Mobile telephones/paging signals
35.48	Special industrial radio service
35.50	Mobile telephones/paging signals
35.52	Special industrial radio service
35.54	Mobile telephones/paging signals
35.56	Paging signals
35.58	Paging signals
35.60	Paging signals
35.62	Mobile telephones/paging signals
35.66	Mobile telephones/paging signals
36.22	Department of Commerce
36.25	Environmental Protection Agency
36.33	Department of Energy
36.40	U.S. Marine Corps
36.65	U.S. Navy
36.71	U.S. Army
36.87	U.S. Army
37.30	Michigan State Police
38.40	U.S. Army
38.69	U.S. Army Corps of Engineers
38.89	U.S. Army Corps of Engineers
39.10	Maryland State Police
39.12	South Dakota State Police

Frequency	Usage	Frequency	Usage
39.26	Maryland State Police	42.44	Massachusetts State Police
39.50	Police departments	42.44	Oregon State Police
39.92	Louisiana State Police	42.46	Nebraska State Police
40.10	National Guard	42.48	Michigan State Police
40.27	Department of Commerce	42.50	Illinois State Police
40.39	Fish and Wildlife Service	42.50	Nevada State Police
40.68	Industrial, scientific, and	42.50	Paging signals [Canada]
	medical devices	42.54	Massachusetts State Police
40.71	Environmental Protection	42.56	Michigan State Police
	Agency	42.56	Oregon State Police
41.00	National Guard	42.56	Tennessee State Police
41.20	U.S. Coast Guard	42.58	Michigan State Police
41.31	Department of Energy	42.64	Connecticut State Police
41.55	U.S. Coast Guard	42.68	Connecticut State Police
41.70	U.S. Marine Corps	42.70	Nevada State Police
41.75	U.S. Coast Guard	42.74	Nevada State Police
41.80	Drug Enforcement	42.78	Oregon State Police
	Administration	42.80	Nevada State Police
41.93	U.S. Marine Corps	42.82	North Carolina State Police
42.02	Mississippi State Police	42.98	Business radio service [low
42.02	Missouri State Police		power operations]
42.04	Nebraska State Police	43.02	Special industrial radio
42.06	South Carolina State Police		service
42.10	West Virginia State Police	43.04	Special industrial radio
42.12	Indiana State Police		service
42.14	New York State Police	43.06	Special industrial radio
42.16	Indiana State Police		service
42.18	California State Police	43.08	Special industrial radio
42.20	Connecticut State Police		service
42.20	Indiana State Police	43.10	Special industrial radio
42.20	Mississippi State Police		service
42.20	Nebraska State Police	43.12	Special industrial radio
42.24	Connecticut State Police		service
42.26	Indiana State Police	43.14	Special industrial radio
42.26	West Virginia State Police		service
42.30	Connecticut State Police	43.20	Paging signals
42.34	California State Police	43.22	Paging signals
42.34	Massachusetts State Police	43.24	Paging signals
42.36	Tennessee State Police	43.26	Paging signals
42.38	Massachusetts State Police	43.28	Special industrial radio
42.42	Massachusetts State Police		service
43.42	Tennessee State Police	43.30	Paging signals

Frequency Usage

43.32	Special industrial radio service
43.34	Paging signals
43.36	Special industrial radio service
43.38	Paging signals
43.40	Special industrial radio service
43.42	Paging signals
43.44	Special industrial radio service
43.46	Paging signals
43.48	Special industrial radio service
43.50	Paging signals
43.52	Special industrial radio service
43.54	Paging signals
43.56	Paging signals
43.58	Paging signals
43.60	Paging signals
43.62	Paging signals
43.66	Paging signals
44.70	Oklahoma State Police
44.78	Arkansas State Police
44.82	New Hampshire State Police
45.02	Ohio State Police
45.14	Kansas State Police
45.18	Kansas State Police
45.18	New Hampshire State Police
45.22	Oklahoma State Police
45.46	Michigan State Police
45.50	Michigan State Police
46.06	Fire departments
46.60	U.S. Navy [reserve units]
46.61	Cordless telephone base units
46.63	Cordless telephone base units
46.65	National Guard
46.67	Cordless telephone base units

Frequency Usage

46.71	Cordless telephone base units
46.73	Cordless telephone base units
46.77	Cordless telephone base units
46.83	Cordless telephone base units
46.87	Cordless telephone base units
46.90	National Guard
46.93	Cordless telephone base units
46.97	Cordless telephone base units
47.42	Red Cross
47.46	Red Cross/search and rescue operations
47.50	Red Cross
48.56	Petroleum radio service
48.58	Petroleum radio service
48.60	Petroleum radio service
48.62	Petroleum radio service
49.50	Petroleum radio service
49.52	Special industrial radio service
49.54	Special industrial radio service
49.58	Special industrial radio service
49.66	Red Cross
49.67	Walkie-talkies and room monitors/cordless telephone handsets
49.77	Walkie-talkies and room monitors/cordless telephone handsets
49.80	Paging signals [Canada]
49.80	U.S. Navy
49.83	Walkie-talkies and room monitors/cordless telephone handsets

Frequency	*Usage*
49.845	Walkie-talkies and room monitors/cordless telephone handsets
49.86	Walkie-talkies and room monitors/cordless telephone handsets
49.875	Walkie-talkies and room monitors/cordless telephone handsets
49.89	Walkie-talkies and room monitors/cordless telephone handsets
49.93	Walkie-talkies and room monitors/cordless telephone handsets
49.97	Walkie-talkies and room monitors/cordless telephone handsets
49.99	Walkie-talkies and room monitors/cordless telephone handsets
52.525	Simplex ham radio communications
53.01	Ham radio repeater output channel
53.03	Ham radio repeater output channel
53.05	Ham radio repeater output channel
53.07	Ham radio repeater output channel
53.09	Ham radio repeater output channel
53.11	Ham radio repeater output channel
53.13	Ham radio repeater output channel
52.15	Ham radio repeater output channel
53.17	Ham radio repeater output channel
53.19	Ham radio repeater output channel

Frequency	*Usage*
53.21	Ham radio repeater output channel
53.23	Ham radio repeater output channel
53.25	Ham radio repeater output channel
53.27	Ham radio repeater output channel
53.29	Ham radio repeater output channel
53.31	Ham radio repeater output channel
53.33	Ham radio repeater output channel
53.35	Ham radio repeater output channel
53.37	Ham radio repeater output channel
53.39	Ham radio repeater output channel
53.41	Ham radio repeater output channel
53.43	Ham radio repeater output channel
53.45	Ham radio repeater output channel
53.47	Ham radio repeater output channel
53.55	Ham radio repeater output channel
53.57	Ham radio repeater output channel
53.59	Ham radio repeater output channel
53.61	Ham radio repeater output channel
53.63	Ham radio repeater output channel
53.65	Ham radio repeater output channel
53.67	Ham radio repeater output channel
53.69	Ham radio repeater output channel

Frequency	Usage		Frequency	Usage
53.71	Ham radio repeater output channel		72.11	Remote control
			72.13	Remote control
53.73	Ham radio repeater output channel		72.15	Remote control
			72.17	Remote control
53.75	Ham radio repeater output channel		72.19	Remote control
			72.21	Remote control
53.77	Ham radio repeater output channel		72.23	Remote control
			72.25	Remote control
53.79	Ham radio repeater output channel		72.27	Remote control
			72.29	Remote control
53.81	Ham radio repeater output channel		72.30	U.S. Air Force
			72.31	Remote control
53.83	Ham radio repeater output channel		72.33	Remote control
			72.35	Remote control
53.85	Ham radio repeater output channel		72.37	Remote control
			72.39	Remote control
53.87	Ham radio repeater output channel		72.41	Remote control
			72.43	Remote control
53.89	Ham radio repeater output channel		72.45	Remote control
			72.47	Remote control
53.91	Ham radio repeater output channel		72.49	Remote control
			72.51	Remote control
53.93	Ham radio repeater output channel		72.53	Remote control
			72.55	Remote control
53.95	Ham radio repeater output channel		72.57	Remote control
			72.59	Remote control
53.97	Ham radio repeater output channel		72.61	Remote control
			72.63	Remote control
53.99	Ham radio repeater output channel		72.65	Remote control
56.50	U.S. Army		72.67	Remote control
66.90	U.S. Air Force [Thunderbirds precision flying team]		72.69	Remote control
			72.70	U.S. Air Force
			72.71	Remote control
72.00	U.S. Army		72.73	Remote control
72.01	Remote control		72.75	Remote control
72.02	Petroleum radio service		72.77	Remote control
72.03	Remote control		72.79	Remote control
72.04	Petroleum radio service		72.81	Remote control
72.05	Remote control		72.83	Remote control
72.06	Petroleum radio service		72.85	Remote control
72.07	Remote control		72.87	Remote control
72.08	Petroleum radio service		72.89	Remote control
72.09	Remote control			

Frequency	Usage	Frequency	Usage
72.91	Remote control	75.39	Wireless microphones/ remote control
72.93	Remote control	75.41	Remote control
72.95	Remote control	75.42	Petroleum radio service
72.97	Remote control	75.43	Remote control
72.99	Remote control	75.44	Petroleum radio service
74.61	Wireless microphones/ remote control	75.45	Remote control
74.63	Wireless microphones/ remote control	75.46	Petroleum radio service
		75.47	Remote control
74.65	Wireless microphones/ remote control	75.48	Petroleum radio service
		75.49	Remote control
74.67	Wireless microphones/ remote control	75.50	Petroleum radio service
		75.51	Remote control
74.69	Wireless microphones/ remote control	75.52	Petroleum radio service
		75.53	Remote control
74.71	Wireless microphones/ remote control	75.55	Remote control
		75.57	Remote control
74.73	Wireless microphones/ remote control	75.59	Remote control
		75.61	Remote control
74.75	Wireless microphones/ remote control	75.63	Remote control
		75.65	Remote control
74.77	Wireless microphones/ remote control	75.67	Remote control
		75.69	Remote control
74.79	Wireless microphones/ remote control	75.71	Remote control
		75.73	Remote control
75.21	Wireless microphones/ remote control	75.75	Remote control
		75.77	Remote control
75.23	Wireless microphones/ remote control	75.79	Remote control
		75.81	Remote control
75.25	Wireless microphones/ remote control	75.83	Remote control
		75.85	Remote control
75.27	Wireless microphones/ remote control	75.87	Remote control
		75.89	Remote control
75.29	Wireless microphones/ remote control	75.91	Remote control
		75.93	Remote control
75.31	Wireless microphones/ remote control	75.95	Remote control
		75.97	Remote control
75.33	Wireless microphones/ remote control	75.99	Remote control
		121.50	Aircraft emergencies
75.35	Wireless microphones/ remote control	121.60	Ground control
		121.65	Ground control
75.37	Wireless microphones/ remote control	121.70	Unicom/ground control

Frequency	Usage	Frequency	Usage
121.75	Ground control/Russian manned space missions	123.50	Aircraft flight schools/ balloons/blimps/gliders
121.80	Ground control	126.20	National Guard
121.85	Ground control	126.40	U.S. Coast Guard helicopters
121.90	Ground control		
121.92	Blimps	136.11	Satellites
121.95	Aircraft flight schools/ gliders	136.37	Satellites
		136.38	Satellites
122.00	Flight service stations	136.86	Satellites
122.05	Flight service stations	137.076	Satellites
122.10	Flight service stations	137.08	Satellites
122.15	Flight service stations	137.17	Satellites
122.20	Flight service stations	137.35	Satellites
122.25	Blimps	137.62	Satellites
122.30	Flight service stations	138.575	Federal Emergency Management Agency
122.35	Flight service stations		
122.40	Flight service stations	138.75	U.S. Army [search and rescue operations]
122.45	Flight service stations		
122.50	Flight service stations/ balloons	139.10	National Guard
		139.20	U.S. Army [reserve units]
122.60	Flight service stations	139.825	Federal Emergency Management Agency
122.75	Air to air/air shows/gliders		
122.775	Balloons/blimps/gliders	141.06	Military civil emergency
122.85	Air to air/air shows	141.12	Military civil emergency
122.90	Multicom/air shows/air to air/search and rescue operations	141.465	Military civil emergency
		141.95	Federal Emergency Management Agency
122.925	Air shows/search and rescue operations	142.23	Federal Emergency Management Agency
123.025	Helicopters	142.35	Federal Emergency Management Agency
123.05	Helicopters		
123.075	Helicopters	142.375	Federal Emergency Management Agency
123.10	U.S. Coast Guard helicopters/search and rescue helicopters	142.40	Federal Emergency Management Agency
		142.417	Russian manned space missions [wideband FM]
123.125	Air shows/balloons/blimps		
123.20	Aircraft flight schools	142.425	Federal Emergency Management Agency
123.30	Aircraft flight schools/ balloons/blimps/gliders	142.44	Military common civil emergency channel
123.40	Aircraft flight schools/ balloons/blimps		
123.45	Air to air/air shows	142.975	Federal Emergency Management Agency

Frequency Usage

143.00	Federal Emergency Management Agency
143.625	Russian manned space missions [wideband FM]
145.11	Ham radio repeater output channel
145.13	Ham radio repeater output channel
145.15	Ham radio repeater output channel
145.17	Ham radio repeater output channel
145.19	Ham radio repeater output channel
145.21	Ham radio repeater output channel
145.23	Ham radio repeater output channel
145.25	Ham radio repeater output channel
145.27	Ham radio repeater output channel
145.29	Ham radio repeater output channel
145.31	Ham radio repeater output channel
145.33	Ham radio repeater output channel
145.35	Ham radio repeater output channel
145.37	Ham radio repeater output channel
145.39	Ham radio repeater output channel
145.41	Ham radio repeater output channel
145.43	Ham radio repeater output channel
145.45	Ham radio repeater output channel
145.47	Ham radio repeater output channel
145.49	Ham radio repeater output channel

Frequency Usage

145.55	Ham radio operations aboard the Space Shuttle and Mir
146.52	Simplex ham radio communications
146.55	Simplex ham radio communications
146.58	Simplex ham radio communications
146.61	Ham radio repeater output channel
146.64	Ham radio repeater output channel
146.67	Ham radio repeater output channel
146.70	Ham radio repeater output channel
146.73	Ham radio repeater output channel
146.76	Ham radio repeater output channel
146.79	Ham radio repeater output channel
146.82	Ham radio repeater output channel
146.85	Ham radio repeater output channel
146.88	Ham radio repeater output channel
146.91	Ham radio repeater output channel
146.94	Ham radio repeater output channel
146.97	Ham radio repeater output channel
147.00	Ham radio repeater output channel
147.03	Ham radio repeater output channel
147.06	Ham radio repeater output channel
147.09	Ham radio repeater output channel

Frequency Usage

147.12	Ham radio repeater output channel
147.15	Ham radio repeater output channel
147.18	Ham radio repeater output channel
147.21	Ham radio repeater output channel
147.24	Ham radio repeater output channel
147.27	Ham radio repeater output channel
147.30	Ham radio repeater output channel
147.33	Ham radio repeater output channel
147.36	Ham radio repeater output channel
147.39	Ham radio repeater output channel
148.15	Civil Air Patrol
148.22	National Aeronautics and Space Administration
149.175	U.S. Army [security]
149.245	Search and rescue operations
149.91	U.S. Army [civil emergencies]
151.16	Illinois State Police
151.46	Wisconsin State Police
151.49	Itinerant users
151.505	Special industrial radio service
151.625	Itinerant users
152.03	Mobile telephones/paging signals
152.06	Mobile telephones/paging signals
152.09	Mobile telephones/paging signals
152.12	Mobile telephones/paging signals
152.15	Mobile telephones/paging signals

Frequency Usage

152.18	Mobile telephones/paging signals
152.21	Mobile telephones/paging signals
152.24	Paging signals
152.51	Mobile telephones/paging signals
152.54	Mobile telephones/paging signals
152.57	Mobile telephones/paging signals
152.60	Mobile telephones/paging signals
152.63	Mobile telephones/paging signals
152.66	Mobile telephones/paging services
152.69	Mobile telephones/paging signals
152.72	Mobile telephones/paging signals
152.75	Mobile telephones/paging signals
152.78	Mobile telephones/paging signals
152.81	Mobile telephones/paging signals
152.84	Paging signals
152.87	Motion picture radio service
152.90	Motion picture radio service
152.93	Motion picture radio service
152.96	Motion picture radio service
152.99	Motion picture radio service
153.02	Motion picture radio service
154.28	Fire department intersystem communications

Frequency	Usage
154.57	Business radio service [low power operations]
154.60	Business radio service [low power operations]
154.655	Texas State Police
154.665	Connecticut State Police
154.665	Kentucky State Police
154.665	New York State Police
154.665	Virginia State Police
154.68	Colorado State Police
154.68	Georgia State Police
154.695	Colorado State Police
154.695	Michigan State Police
154.695	Montana State Police
154.695	Rhode Island State Police
154.755	Pennsylvania State Police
154.785	Arkansas State Police
154.785	Wyoming State Police
154.83	Connecticut State Police
154.905	California State Police
154.905	Colorado State Police
154.905	Minnesota State Police
154.905	Missouri State Police
154.905	Oklahoma State Police
154.905	Tennessee State Police
154.92	Alabama State Police
154.92	Florida State Police
154.92	Kansas State Police
154.92	Massachusetts State Police
154.92	Missouri State Police
154.92	Nevada State Police
154.935	Maine State Police
154.95	Texas State Police
155.16	Search and rescue operations
155.175	Emergency medical services
155.235	Search and rescue operations
155.25	Alaska State Police
155.325	Emergency medical services
155.34	Emergency medical services
155.355	Emergency medical services

Frequency	Usage
155.37	Police department systemwide bulletins
155.385	Emergency medical services
155.40	Emergency medical services
155.43	Iowa State Police
155.445	Indiana State Police
155.445	North Carolina State Police
155.445	Rhode Island State Police
155.445	South Carolina State Police
155.445	Wyoming State Police
155.46	Illinois State Police
155.46	New Jersey State Police
155.46	Pennsylvania State Police
155.46	Illinois State Police
155.475	Police intersystem communications [mutual aid]
155.505	Iowa State Police
155.505	Maine State Police
155.505	West Virginia State Police
155.535	Mississippi State Police
155.55	New Mexico State Police
155.73	Maryland State Police
155.745	Utah State Police
155.97	Washington State Police
156.03	North Dakota State Police
156.05	Marine communications
156.09	Arizona State Police
156.09	Maine State Police
156.18	Florida State Police
156.25	Marine communications
156.275	Marine communications
156.30	Marine communications
156.325	Marine communications
156.35	Marine communications
156.375	Marine communications
156.40	Marine communications
156.425	Marine communications
156.45	Marine communications
156.475	Marine communications
156.50	Marine communications
156.525	Marine communications

Frequency Usage

156.55	Marine communications
156.575	Marine communications
156.625	Marine communications
156.675	Marine communications
156.70	Marine communications
156.725	Marine communications
156.80	Marine communications [distress and calling frequency]
156.85	Marine communications
156.875	Marine communications
156.90	Marine communications
156.925	Marine communications
156.95	Marine communications
156.975	Marine communications
157.00	Marine communications
157.025	Marine communications
157.05	Marine communications
157.075	Marine communications
157.10	Marine communications
157.125	Marine communications
157.15	Marine communications
156.175	Marine communications
157.20	Marine communications
157.225	Marine communications
157.25	Marine communications
157.275	Marine communications
157.30	Marine communications
157.325	Marine communications
157.35	Marine communications
157.375	Marine communications
157.40	Marine communications
158.10	Paging signals
158.40	Special industrial radio service
158.70	Paging signals
158.73	Maine State Police
158.79	Alabama State Police
158.91	North Dakota State Police
158.97	Mississippi State Police
159.03	Arizona State Police
159.075	Washington State Police

Frequency Usage

159.09	Texas State Police
159.15	Rhode Island State Police
159.21	North Dakota State Police
159.21	Oklahoma State Police
160.335	Minnesota State Police
161.02	Surveillance devices, state law enforcement agencies
161.62	Surveillance devices, state law enforcement agencies
161.64	Broadcast auxiliary
161.67	Broadcast auxiliary
161.70	Broadcast auxiliary
161.73	Broadcast auxiliary
161.76	Broadcast auxiliary
162.025	National Aeronautics and Space Administration
162.40	Weather broadcasts
162.425	National Weather Service
162.45	National Weather Service
162.475	Weather broadcasts
162.50	National Weather Service
162.525	National Weather Service
162.55	Weather broadcasts
162.825	U.S. Customs Service
163.10	Numerous U.S. government agencies
163.4875	National Guard
163.5125	Military disaster relief
163.85	Numerous U.S. government agencies
163.9125	Federal Bureau of Investigation
164.10	Secret Service
164.30	Department of Health and Human Services
164.45	Environmental Protection Agency
164.65	Secret Service
164.8625	Federal Emergency Management Agency

Frequency	Usage	Frequency	Usage
165.1875	Surveillance devices, state law enforcement agencies	169.55	Wireless microphones
		169.85	U.S. Postal Service
165.205	Department of the Treasury	170.15	Broadcast auxiliary
165.2125	Surveillance devices, state law enforcement agencies	170.175	U.S. Postal Service
		170.245	Wireless microphones
		170.925	U.S. Bureau of Prisons [riots, escapes, and emergencies]
165.375	Bureau of Alcohol, Tobacco, and Firearms/ Secret Service		
		171.575	Minnesota State Police
		171.905	Wireless microphones
165.6625	Federal Emergency Management Agency	172.75	Nuclear Regulatory Commission
165.7875	Secret Service	172.80	Federal Communications Commission
166.00	Russian manned space program telemetry channel		
		173.225	Motion picture radio service/newspapers
166.20	Department of the Treasury	173.275	Motion picture radio service/newspapers
166.2125	Secret Service		
166.25	Broadcast auxiliary	173.325	Motion picture radio service/newspapers
166.4625	Department of the Treasury/ Secret Service/U.S. Customs	173.375	Motion picture radio service/newspapers
166.5125	Secret Service	174.60	Wireless microphones
166.5875	U.S. Customs Service	174.80	Wireless microphones
167.00	Internal Revenue Service	175.00	Wireless microphones
167.05	Federal Communications Commission	177.60	Wireless microphones
		180.60	Wireless microphones
167.10	Internal Revenue Service	181.60	Wireless microphones
167.3575	Surveillance devices, state law enforcement agencies	183.60	Wireless microphones
		184.00	Wireless microphones
		186.60	Wireless microphones
167.5625	Federal Bureau of Investigation	190.60	Wireless microphones
		192.60	Wireless microphones
167.60	Federal Bureau of Investigation	194.60	Wireless microphones
		195.60	Wireless microphones
168.35	National Aeronautics and Space Administration	196.60	Wireless microphones
		199.60	Wireless microphones
169.00	U.S. Postal Service	202.40	Wireless microphones
169.10	Nuclear Regulatory Commission	203.40	Wireless microphones
		222.94	Ham radio repeater output channel
169.385	Surveillance devices, state law enforcement agencies	223.50	Simplex ham radio communications

Frequency Usage

223.72	Ham radio repeater output channel
223.82	Ham radio repeater output channel
223.88	Ham radio repeater output channel
223.94	Ham radio repeater output channel
224.06	Ham radio repeater output channel
224.08	Ham radio repeater output channel
224.12	Ham radio repeater output channel
224.22	Ham radio repeater output channel
224.30	Ham radio repeater output channel
224.40	Ham radio repeater output channel
224.44	Ham radio repeater output channel
224.52	Ham radio repeater output channel
224.60	Ham radio repeater output channel
224.64	Ham radio repeater output channel
224.66	Ham radio repeater output channel
224.78	Ham radio repeater output channel
224.82	Ham radio repeater output channel
224.90	Ham radio repeater output channel
224.94	Ham radio repeater output channel
224.98	Ham radio repeater output channel
225.00	U.S. Navy
229.60	U.S. Army/National Guard
231.00	National Aeronautics and Space Administration

Frequency Usage

235.00	U.S. Navy
235.10	U.S. Air Force [air refueling operations]
236.60	U.S. Air Force [control towers]
237.90	U.S. Coast Guard
238.90	U.S. Air Force [air refueling operations]
239.80	Aviation weather bulletins
240.40	National Aeronautics and Space Administration
240.60	National Aeronautics and Space Administration/ U.S. Navy
240.80	National Aeronautics and Space Administration
241.00	U.S. Army/National Guard helicopters
241.20	National Aeronautics and Space Administration
241.40	U.S. Navy
241.60	National Aeronautics and Space Administration
241.80	National Aeronautics and Space Administration
242.00	National Aeronautics and Space Administration
242.20	U.S. Air Force/Tactical Air Command
243.00	Aircraft emergencies
245.70	Royal Canadian Air Force
250.80	U.S. Navy
251.60	U.S. Navy
255.40	Federal Aviation Administration flight services
256.20	National Aeronautics and Space Administration
257.80	U.S. Air Force [control towers]
259.70	Space Shuttle
260.20	U.S. Air Force [air refueling operations]

Frequency	Usage
264.20	U.S. Navy
270.60	U.S. Navy
272.70	Federal Aviation Administration flight services
273.50	U.S. Air Force
273.80	U.S. Air Force/Strategic Air Command
275.10	U.S. Navy/U.S. Coast Guard
275.80	U.S. Air Force [control towers]
276.50	U.S. Air Force [air refueling operations]
276.90	U.S. Air Force
277.00	U.S. Navy
277.80	U.S. Coast Guard/U.S. Navy
279.00	Space Shuttle
280.50	U.S. Air Force/Tactical Air Command
282.50	U.S. Air Force/Tactical Air Command
282.70	U.S. Air Force [air refueling operations]
282.80	U.S. Coast Guard [helicopters and search and rescue operations]
283.50	U.S. Air Force
283.70	U.S. Air Force/Tactical Air Command
285.00	U.S. Navy/U.S. Coast Guard
285.40	U.S. Air Force [control towers]
286.90	U.S. Air Force
289.40	U.S. Air Force
289.50	U.S. Air Force
289.70	U.S. Air Force [air refueling operations]
289.80	U.S. Navy
289.90	U.S. Air Force
292.10	U.S. Air Force/Tactical Air Command

Frequency	Usage
293.00	U.S. Air Force [air refueling operations]
294.20	U.S. Air Force
295.40	U.S. Air Force [air refueling operations]
295.70	U.S. Air Force
295.80	U.S. Air Force [air refueling operations]
296.20	U.S. Air Force
296.80	Space Shuttle
297.00	U.S. Air Force/Military Airlift Command
300.60	U.S. Navy [air to air]
301.00	U.S. Navy
302.70	U.S. Air Force
305.40	U.S. Air Force
305.50	U.S. Air Force
305.60	U.S. Air Force/Tactical Air Command
305.70	U.S. Air Force/Tactical Air Command
306.60	U.S. Air Force/Tactical Air Command
307.70	U.S. Navy
311.00	U.S. Air Force/Strategic Air Command
313.60	U.S. Air Force/Tactical Air Command
316.50	Royal Canadian Air Force
318.30	U.S. Air Force
319.40	U.S. Air Force/Military Airlift Command
319.50	U.S. Air Force [air refueling operations]
320.20	U.S. Navy
320.90	U.S. Air Force [air refueling operations]
321.00	U.S. Air Force/Strategic Air Command
321.20	U.S. Air Force/Strategic Air Command
322.60	U.S. Air Force
326.30	U.S. Air Force

Frequency	Usage	Frequency	Usage
335.70	U.S. Air Force	375.20	Aviation weather bulletins
335.80	U.S. Air Force	375.70	U.S. Air Force/Strategic Air Command
336.10	U.S. Air Force		
338.50	U.S. Air Force	376.20	U.S. Air Force/Tactical Air Command
340.20	U.S. Navy [control towers]		
340.60	U.S. Air Force	378.10	U.S. Air Force
340.80	U.S. Air Force/Military Airlift Command	378.40	U.S. Air Force/Tactical Air Command
342.20	U.S. Air Force/Strategic Air Command	378.80	U.S. Air Force
		381.30	U.S. Air Force/Tactical Air Command
342.50	Aviation weather bulletins		
343.00	U.S. Air Force/Tactical Air Command	381.70	U.S. Coast Guard helicopters
343.50	U.S. Air Force/Tactical Air Command	381.80	U.S. Coast Guard helicopters
344.60	Aviation weather bulletins	382.50	U.S. Air Force
344.70	U.S. Air Force [air refueling operations]	382.90	U.S. Air Force
		383.90	U.S. Coast Guard
345.00	U.S. Navy	384.40	U.S. Navy
348.60	U.S. Air Force [control towers]	384.80	U.S. Air Force
		385.00	U.S. Navy
349.40	U.S. Air Force/Tactical Air Command	387.40	U.S. Navy
		387.90	U.S. Air Force/Tactical Air Command
354.20	U.S. Air Force		
355.00	U.S. Navy	390.90	U.S. Air Force/Military Airlift Command
357.70	U.S. Air Force		
357.90	U.S. Navy	391.90	U.S. Navy
358.90	U.S. Navy	395.90	U.S. Navy
359.40	U.S. Navy	398.50	U.S. Air Force
360.20	U.S. Navy [control towers]	406.325	U.S. Postal Service
360.40	U.S. Navy	408.05	Numerous U.S. government agencies
363.80	U.S. Air Force [control towers]	409.20	Interstate Commerce Commission
364.20	North American Air Defense Command (NORAD)	409.275	U.S. Postal Service
		413.925	Federal Reserve Banks
364.60	U.S. Air Force	414.70	Internal Revenue Service
368.60	U.S. Air Force	415.00	Internal Revenue Service
369.10	U.S. Air Force	415.05	U.S. Postal Service
370.40	U.S. Air Force	415.20	General Services Administration
372.20	U.S. Air Force		
372.80	U.S. Air Force/Military Airlift Command	415.70	Air Force One [telephone calls]

Frequency	Usage	Frequency	Usage
416.175	U.S. Attorney offices	446.70	Ham radio repeater output channel
418.05	Numerous U.S. government agencies	447.00	Ham radio repeater output channel
418.10	U.S. Postal Service	448.50	Ham radio repeater output channel
441.45	Ham radio repeater output channel	449.00	Ham radio repeater output channel
441.95	Ham radio repeater output channel	449.50	Ham radio repeater output channel
442.10	Ham radio repeater output channel	449.60	Ham radio repeater output channel
442.20	Ham radio repeater output channel	449.75	Ham radio repeater output channel
442.45	Ham radio repeater output channel	450.05	Broadcast auxiliary
442.80	Ham radio repeater output channel	450.0875	Broadcast auxiliary
443.00	Ham radio repeater output channel	450.10	Broadcast auxiliary
		450.1125	Broadcast auxiliary
443.15	Ham radio repeater output channel	450.1375	Broadcast auxiliary
		450.15	Broadcast auxiliary
443.45	Ham radio repeater output channel	450.1625	Broadcast auxiliary
		450.1875	Broadcast auxiliary
443.60	Ham radio repeater output channel	450.20	Broadcast auxiliary
		450.2125	Broadcast auxiliary
443.85	Ham radio repeater output channel	450.2375	Broadcast auxiliary
		450.25	Broadcast auxiliary
444.00	Ham radio repeater output channel	450.2625	Broadcast auxiliary
		450.2875	Broadcast auxiliary
444.05	Ham radio repeater output channel	450.30	Broadcast auxiliary
		450.3125	Broadcast auxiliary
444.20	Ham radio repeater output channel	450.3375	Broadcast auxiliary
		450.35	Broadcast auxiliary
444.38	Ham radio repeater output channel	450.3625	Broadcast auxiliary
		450.3875	Broadcast auxiliary
444.50	Ham radio repeater output channel	450.40	Broadcast auxiliary
444.65	Ham radio repeater output channel	450.4125	Broadcast auxiliary
		450.4375	Broadcast auxiliary
444.80	Ham radio repeater output channel	450.45	Broadcast auxiliary
		450.4625	Broadcast auxiliary
444.95	Ham radio repeater output channel	450.4875	Broadcast auxiliary
		450.50	Broadcast auxiliary
446.00	Simplex ham radio communications	450.5125	Broadcast auxiliary
		450.5375	Broadcast auxiliary

Frequency	Usage	Frequency	Usage
450.55	Broadcast auxiliary	454.225	Mobile telephones/paging signals
450.5625	Broadcast auxiliary		
450.5875	Broadcast auxiliary	454.25	Mobile telephones/paging signals
450.60	Broadcast auxiliary		
450.6125	Broadcast auxiliary	454.275	Mobile telephones/paging signals
450.65	Broadcast auxiliary		
450.70	Broadcast auxiliary	454.30	Mobile telephones/paging signals
450.75	Broadcast auxiliary		
450.80	Broadcast auxiliary	454.325	Mobile telephones/paging signals
450.85	Broadcast auxiliary		
450.90	Broadcast auxiliary	454.35	Mobile telephones/paging signals
450.925	Broadcast auxiliary		
451.80	Special industrial radio service	454.375	Mobile telephones/paging signals
452.975	Newspapers	454.40	Mobile telephones/paging signals
453.00	Newspapers		
453.25	Minnesota State Police	454.425	Mobile telephones/paging signals
453.275	North Dakota State Police	454.45	Mobile telephones/paging signals
453.30	Kentucky State Police		
453.35	Virginia State Police	454.475	Mobile telephones/paging signals
453.375	South Dakota State Police		
453.45	Louisiana State Police	454.50	Mobile telephones/paging signals
453.45	North Dakota State Police		
454.475	Washington State Police	454.525	Mobile telephones/paging signals
453.625	Iowa State Police		
453.925	Washington State Police	454.55	Mobile telephones/paging signals
454.025	Mobile telephones/paging signals	454.575	Mobile telephones/paging signals
454.05	Mobile telephones/paging signals	454.60	Mobile telephones/paging signals
454.075	Mobile telephones/paging signals	454.625	Mobile telephones/paging signals
454.10	Mobile telephones/paging signals	454.65	Mobile telephones/paging signals
454.125	Mobile telephones/paging signals	455.05	Broadcast auxiliary
454.15	Mobile telephones/paging signals	455.0875	Broadcast auxiliary
		455.10	Broadcast auxiliary
454.175	Mobile telephones/paging signals	455.1125	Broadcast auxiliary
		455.1375	Broadcast auxiliary
454.20	Mobile telephones/paging signals	455.15	Broadcast auxiliary
		455.1625	Broadcast auxiliary

Frequency	Usage	Frequency	Usage
455.1875	Broadcast auxiliary	457.60	Business radio service [low power operations]
455.20	Broadcast auxiliary		
455.2125	Broadcast auxiliary	458.25	Minnesota State Police
455.2375	Broadcast auxiliary	458.35	Virginia State Police
455.25	Broadcast auxiliary	458.4875	Georgia State Police
455.2625	Broadcast auxiliary	460.025	Vermont State Police
455.2875	Broadcast auxiliary	460.125	West Virginia State Police
455.30	Broadcast auxiliary	460.15	New Mexico State Police
455.3125	Broadcast auxiliary	460.225	Arizona State Police
455.3375	Broadcast auxiliary	460.225	Maine State Police
455.35	Broadcast auxiliary	460.275	Idaho State Police
455.3625	Broadcast auxiliary	460.525	Nebraska State Police
455.3875	Broadcast auxiliary	460.50	Delaware State Police
455.40	Broadcast auxiliary	460.50	Vermont State Police
455.4125	Broadcast auxiliary	462.55	General mobile radio service
455.4375	Broadcast auxiliary		
455.45	Broadcast auxiliary	462.5625	General mobile radio service
455.4625	Broadcast auxiliary		
455.4875	Broadcast auxiliary	462.575	General mobile radio service
455.50	Broadcast auxiliary		
455.5125	Broadcast auxiliary	462.5875	General mobile radio service
455.5375	Broadcast auxiliary		
455.55	Broadcast auxiliary	462.60	General mobile radio service
455.5625	Broadcast auxiliary		
455.5825	Broadcast auxiliary	462.6125	General mobile radio service
455.60	Broadcast auxiliary		
455.6125	Broadcast auxiliary	462.625	General mobile radio service
455.65	Broadcast auxiliary		
455.70	Broadcast auxiliary	462.6375	General mobile radio service
455.75	Broadcast auxiliary		
455.80	Broadcast auxiliary	462.65	General mobile radio service
455.85	Broadcast auxiliary		
455.90	Broadcast auxiliary	462.6625	General mobile radio service
455.925	Broadcast auxiliary		
456.80	Special industrial radio service	462.675	General mobile radio service
		462.6875	General mobile radio service
457.525	Business radio service [low power operations]		
		462.70	General mobile radio service
457.55	Business radio service [low power operations]		
		462.725	General mobile radio service
457.575	Business radio service [low power operations]		
		464.40	Idaho State Police

Frequency	Usage	Frequency	Usage
464.50	Itinerant users	931.0625	Paging signals
464.55	Itinerant users	931.0875	Paging signals
465.0125	Oklahoma State Police	931.1125	Paging signals
465.125	Wisconsin State Police	931.1375	Paging signals
465.15	New Mexico State Police	931.1625	Paging signals
465.1625	Florida State Police	931.1875	Paging signals
465.1625	Oklahoma State Police	931.2125	Paging signals
465.375	Ohio State Police	931.2375	Paging signals
465.3875	Oklahoma State Police	931.2625	Paging signals
465.425	Ohio State Police	931.2875	Paging signals
465.475	Delaware State Police	931.3125	Paging signals
465.525	Ohio State Police	931.3375	Paging signals
465.535	Nebraska State Police	931.3625	Paging signals
465.55	Ohio State Police	931.3875	Paging signals
465.5625	Oklahoma State Police	931.4125	Paging signals
467.75	Business radio service [low power operations]	931.4375	Paging signals
467.775	Business radio service [low power operations]	931.4625	Paging signals
		931.4875	Paging signals
467.80	Business radio service [low power operations]	931.5125	Paging signals
		931.5375	Paging signals
467.825	Business radio service [low power operations]	931.5625	Paging signals
		931.5875	Paging signals
467.85	Business radio service [low power operations]	931.6125	Paging signals
		931.6375	Paging signals
467.875	Business radio service [low power operations]	931.6625	Paging signals
		931.6875	Paging signals
467.90	Business radio service [low power operations]	931.7125	Paging signals
		931.7375	Paging signals
467.925	Business radio service [low power operations]	931.7625	Paging signals
		931.7875	Paging signals
469.50	Itinerant users	931.8125	Paging signals
469.55	Itinerant users	931.8375	Paging signals
906.50	Simplex ham radio communications	931.8625	Paging signals
		931.8875	Paging signals
915.00	Industrial, scientific, and medical devices	931.9125	Paging signals
		931.9375	Paging signals
931.0125	Paging signals	931.9625	Paging signals
931.0375	Paging signals	931.9875	Paging signals

Index